Fundamental
Phenomena
in the
Materials Sciences

Volume 3
Surface Phenomena

Fundamental Phenomena in the Materials Sciences

Volume 3
Surface Phenomena

Proceedings of the Third Symposium on Fundamental Phenomena in the Materials Sciences

Held January 25-26, 1965, at Boston, Mass.

Edited by
L. J. Bonis
Ilikon Corporation
Natick, Massachusetts

P. L. de Bruyn
Department of Metallurgy
Massachusetts Institute of Technology
Cambridge, Massachusetts

and

J. J. Duga
Battelle Memorial Institute
Columbus, Ohio

℗ Springer Science+Business Media, LLC 1966

Library of Congress Catalog Card No. 64-20752

ISBN 978-1-4899-6173-0   ISBN 978-1-4899-6347-5 (eBook)
DOI 10.1007/978-1-4899-6347-5

© *1966 Springer Science+Business Media New York*
Originally published by Plenum Press in 1966
Softcover reprint of the hardcover 1st edition 1966

Foreword

This volume explores in detail the four interrelated branches of the study of surface phenomena—surface thermodynamics, nucleation, diffusion, and fine-particles technology—providing an unusual and comprehensive body of knowledge that will be of interest and practical value to both materials researchers and practising engineers.

The growing awareness—since the advent of the space age—among solid-state physicists, metallurgists, ceramists, chemical engineers, and mechanical engineers of the need for a broad interdisciplinary understanding of the fundamental phenomena common to all materials has led in recent years to the development of a new field of scientific investigation, Materials Science. To help promote interest in and contributions to this new technology, annual symposia on "Fundamental Phenomena in the Materials Sciences" have been organized by the Ilikon Corporation. The first symposium, reported in Volume 1 of this series, was held in Boston, Massachusetts, on February 1 and 2, 1963; sintering and plastic deformation were the main topics of discussion. The second meeting, also held in Boston, on January 27 and 28, 1964, was exclusively concerned with the general interdisciplinary problems related to surface phenomena, that is, all of those physical and chemical areas that are pertinent to the surface of a solid, or to the interface between a solid and a gas, a solid and a liquid, or a solid and a solid. Ten noted researchers, from major universities and industrial organizations, presented papers that dealt with the principles of surface physics and surface chemistry. Surface phenomena were treated, therefore, in a broad sense, with emphasis on structures, electronic configurations, and properties of clean real surfaces, rather than on the generic bases of specific material types.

Although unique and broadly rewarding, that second meeting barely "scratched the surface" of what all the participants at the meeting recognized to be a truly fundamental materials technology subject. It was decided, therefore, to continue these surface phenomena investigations at a third meeting. The papers presented at that conference, held January 25 and 26, 1965, in Boston, together with the panel discussions as they occurred, are presented in this volume.

In the first chapter, P. L. de Bruyn (Massachusetts Institute of Technology) reviews Gibbs' classical thermodynamics treatment of surfaces and interfaces, giving particular attention to Gibbs' concept of a dividing surface. In this context of the classical thermodynamics equations, the author proceeds in his discussion to point out that, in most thermodynamic treatments of multiphases, it is assumed that, when the main interest is in the bulk properties, the surface may be completely ignored. Next, he suggests that for the study of the interface between two juxtaposed contiguous phases it is convenient to introduce a mathematical "dividing" surface, with respect to which all "excess" surface properties may be defined. This dividing surface and the surface properties defined by it, however, must take into account the fact that there is a definite gradient in properties in the vicinity of any interface. Following a discussion of the implications of this gradient, the author defines and discusses surface tension, arguing that it is always directly proportional to the pressure differences across the interface boundary.

In the second chapter, E. W. Hart (General Electric Research Laboratory) continues the consideration of surface tension by describing the results of the application of the recently developed thermodynamic theory for equilibrium in homogeneous systems to the problem of the surface tension of the interface between two fluid phases. He states that the results of the experimentation were inconsistent with earlier conclusions regarding both the concept of surface tension and the existence of minimum surface tension, proving rather that the surface tension of a planar interface was given solely by the difference of two quantities that are characteristic of the two homogeneous bulk phases.

The subject of nucleation is introduced in the third chapter by K. C. Russell (Massachusetts Institute of Technology), who points out that at the present time there is a 1-to-7 order-of-magnitude difference between the experimentally determined and the theoretically predicted behavior of homogeneous nucleation and, furthermore, that the best this situation can be improved is by about one order of magnitude—unless more sophisticated techniques are employed. He describes such techniques as those that would focus attention on those additional energy and entropy characteristics that may be meaningful.

In the next chapter, J. P. Hirth and K. L. Moazed (Ohio State University) describe different nucleation mechanisms related to deposition onto substrates. They demonstrate how through a minimization of the free energy of formation it is possible to determine, in order, the critical nucleus size above which crystal growth can be expected to begin, the volume density of the critical-size nucleus, and the observable nucleation rate. Following this enumeration, the authors describe mechanisms that are dependent principally on the temperature of the substrate. In this chapter they also discuss surface imperfections and the phenomenon of epitaxy.

F. P. Price (General Electric Research Laboratory) discusses, in the fifth chapter, the nucleation and growth of single-crystal organic polymers, presenting evidence that classical nucleation theory applies equally well to polymer crystals and low-molecular-weight materials. He shows that the thermal history of organic solids controls the size of crystalline spherulites and that for polychlorotrifluoroethylene there is correlation between spherulite size and fatigue life.

In the sixth chapter, P. G. Shewmon (Carnegie Institute of Technology) introduces the subject of surface diffusion by examining the experimental techniques used to determine the surface diffusion coefficient D_s. Then he argues that surface diffusion rather than volume is the major factor in the process of fine powder sintering. He describes recent computer studies in support of this minority opinion.

Consideration of the relation of surface diffusion to the sintering process is continued in the seventh chapter, where C. E. Birchenall and J. M. Williams (University of Delaware) express the view that any consideration of surface diffusion must take into account the additional factors of surface impurity size and distribution, crystalline anisotropies, divacancy migration through the surface, the long mean free paths at high temperatures, and the chemical effects that "pin" atoms to the adsorbed species. They discuss these factors in some detail and then proceed to comment on the sintering process itself, expressing the opinion—contrary to the usual—that neck size is more important than particle size.

In the eighth chapter, after commenting on the formation of surface point defects on ionic crystals, authors C. Y. Li and J. M. Blakely (Cornell University) question the legitimacy of utilizing the bulk dielectric constant in the neighborhood of a surface, pointing out that a correction in this constant would make it more difficult to create a surface pair, for any such correction would result in an increase of the polarization component of the removal energy.

In the discussion of surface-initiated failures in structural materials presented in the ninth chapter, I. R. Kramer (Martin Company) points out that the surface of any material has a significant effect on the mechanical behavior of the material and that, in fact, suitable changes in the surface can result in the altering of all of the usual mechanical properties of the material—tensile behavior, fatigue, creep, and stress rupture.

Ultrafine particles in gases is the subject of the final chapter. One of the principal suggestions in this broad, comprehensive discussion by J. Turkevich (Princeton University) is that the formation as well as the texture of these particles are the result of a degree of "memory" residing in an organized aggregate of the materials. Another suggestion is that in the growth of these particles there are none of the growth-promoting driving forces found in crystallites and that these particles are so nearly perfect that the "growth sites" associated with dislocations do not exist. In addition to these ideas, the author describes in detail some of the problems associated with the formation of these ultrafine particles.

Great efforts were put into trying to preserve the flavor of the meeting in this book, examples of which are shown in the two sections where selected discussions are presented.

It is not without pride that we view the ever-mounting interest in these meetings, viewed not only from the number of people attending (because of this, attendance is now limited to "by invitation only"), but also from the high caliber of the participants.

L. J. Bonis

Contents

Some Aspects of
Classical Surface Thermodynamics

P. L. de Bruyn

Department of Metallurgy
Massachusetts Institute of Technology
Cambridge, Massachusetts

INTRODUCTION

The special contribution of matter lying in and around phase boundaries to the total energy of heterogeneous systems is normally ignored in thermodynamic treatises. This neglect implies that the density of energy of a given phase remains uniform up to a mathematical surface separating it from contiguous phases. However, because of the finite, although short, range of action of atomic forces, the assumed sharp phase boundaries should actually be replaced by an interphasal region of finite thickness across which the density of energy or of any other thermodynamic property changes much less abruptly. It is reasonable to expect the normal thickness of this transition region to be of the order of a few molecular diameters; unless the system has a relatively high surface to volume ratio, as would be the case in colloidal systems, the additional contribution of the "surface" atoms to the total energy content may justifiably be ignored.

However, a solid or liquid surface, no matter how small in extent, will be the seat of special physical and chemical phenomena because those atoms or molecules which terminate the condensed phase are subjected to unsymmetrical forces. Adsorption from the gaseous or liquid phase onto solid surfaces, wetting and spreading of liquids on solids, and capillarity are examples of industrially important surface phenomena. Surfaces also play an important, although not dominant, role in such physical and chemical processes as phase transformations, nucleation and growth, electrode reactions, sintering, and electron emission—to name but a few. The colloid chemist has long been aware of the role of surfaces in colloidal phenomena because he has

1

always been concerned with systems of large surface area. In the last few decades, the physicist and the metallurgist have given special attention to surfaces because of increased interest in thin films and precipitation and ageing phenomena in alloy systems.

J. Willard Gibbs may rightfully be called the father of surface science—that branch of science concerned with an understanding of the physical and chemical nature of surfaces and with the formulation of the general principles governing surface phenomena. Surface thermodynamics first took form with the appearance of Gibbs' monumental treatise, "Influence of Surfaces of Discontinuity upon the Equilibrium of Heterogeneous Masses"[6]. It is an everlasting tribute to the genius of this master of thermodynamics that his rigorous and exact exposition still stands unchallenged. A new field of study has developed from many a footnote in this treatise. Gibbs not only anticipated future developments in surface science, but also warned of the pitfalls to be encountered and indicated how they might be avoided. Unfortunately, his warnings have often been ignored by the unwary investigator.

It is the purpose of this paper to review some aspects of surface thermodynamics as originally developed by Gibbs and thereby to provide the point of departure for subsequent discussions of the role of surfaces in materials science. Before proceeding with this task, it might be worthwhile to examine the advances which have been made in surface thermodynamics since Gibbs' treatise. A vast amount of experimental material has been accumulated and analyzed by application of Gibbs' theoretical treatment, but the only significant departure from the thermodynamic approach of Gibbs has been the application of continuum mechanical principles to the description of the transition region. This new approach originated at the end of the nineteenth century with van der Waals [1] and Bakker [2], and it was revived recently by Cahn and Hilliard [3,4] and by Hart [5]. The advantage of this approach to the thermodynamic description of nucleation has been demonstrated by Cahn and Hilliard.

FUNDAMENTAL THERMODYNAMIC RELATIONSHIPS OF HETEROGENEOUS SYSTEMS

This discussion of surface thermodynamics can be introduced by reiterating some useful thermodynamic relations of heterogeneous systems. For a reversible change in state of an open system, the total differential of internal energy dU is given by the relation

$$dU = T\,dS - p\,dV + \sum \mu_i\,dn_i \tag{1}$$

where the extensive quantities S, V, and n_i are the entropy, volume, and number of moles of the ith component of the system, respectively, and the intensive properties T, p, and μ_i are the absolute temperature, pressure, and chemical potential, respectively. The most useful definition of chemical potential is the following one that relates it to the partial molal free energy $\bar{G}_i$:

$$\mu_i = \left(\frac{\partial G}{\partial n_i}\right)_{T,p,n_j} \equiv \bar{G}_i \tag{2}$$

where

$$G \equiv U + pV - TS \tag{3}$$

is the Gibbs free energy of the system and n_j refers to the number of moles of all components except i. The composition of a phase is commonly described by the mole fraction of each component. Mole fractions are pure numbers satisfying the identity

$$\sum x_i = 1$$

where

$$x_i = \frac{n_i}{\sum n_i}$$

The dependence of chemical potential on the temperature, pressure, and composition of a homogeneous phase is expressed by the relation

$$d\mu_i = -\bar{S}_i\,dT + \bar{V}_i\,dp + \sum_{k=1}^{r} \left(\frac{\partial \mu_i}{\partial x_k}\right)_{p,T,x_j \neq x_k} dx_k \tag{4}$$

where $\bar{S}_i$ and $\bar{V}_i$ are the partial molal entropy and volume of species i, respectively. At constant temperature and pressure,

$$\mu_i = \mu_i^0(p, T) + RT \ln \lambda_i x_i \tag{5}$$

where μ_i^0 is the chemical potential of the ith component in its standard state of unit activity and λ_i is a dimensionless activity coefficient which is a function of temperature, pressure, and composition. For very dilute solutions, with component i the solute species, the relation

$$\frac{\partial \mu_i}{\partial x_i} = \frac{RT}{x_i} \tag{6}$$

may be assumed to hold.

Since internal energy is a homogeneous function of the first degree in S, V, and n, integration of equation (1) at constant pressure, temperature, and composition shows that

$$U = TS - pV + \sum_{i=1}^{r} \mu_i n_i \tag{7}$$

We introduce the state function

$$\Omega \equiv U - TS - \sum \mu_i n_i$$
$$\Omega = -pV \tag{8}$$

which may be compared to the well-known free energy functions F and G. We note that, in contrast to the functions F and G, this free energy function does not depend on the choice of a standard state. Differentiation of equation (8) and comparison with equation (1) shows that

$$d\Omega = -S\, dT - p\, dV - \sum n_i\, d\mu_i \tag{9}$$

In turn, differentiation of equation (8) and comparison with equation (9) yields the well-known Gibbs–Duhem relation

$$-S\, dT + V\, dp - \sum n_i\, d\mu_i = 0 \tag{10}$$

which, at constant temperature and pressure, reduces to

$$\sum n_i\, d\mu_i = 0 \tag{11a}$$

or, with the introduction of mole fractions,

$$\sum x_i\, d\mu_i = 0 \tag{11b}$$

When a system comprises two or more homogeneous phases, it also includes one or more physical boundaries. The homogeneous phases are regions of the system in which the thermodynamic properties are independent of position, whereas, at the physical boundaries, most thermodynamic quantities characterizing the state of the system are normally assumed to change abruptly. The condition for internal thermodynamic equilibrium in heterogeneous systems requires that the temperature and pressure be constant throughout and that the chemical potential of each component be the same in all the phases in which it is present, i.e.,

$$\mu_i^\alpha = \mu_i^\beta = \cdots = \mu_i^\xi \tag{12}$$

where α, β, and ξ refer to homogeneous phases. The latter condition also implies that the chemical potential of a possible, but not actual, component of any phase be not less than its value in the phase in which it is present. For a two-phase system at constant temperature and pressure, according to equation (12), the Gibbs–Duhem relation may be written as follows:

$$\sum (n_i^\alpha - n_i^\beta)\, d\mu_i = 0 \tag{13}$$

THE DIVIDING SURFACE AND EXCESS PROPERTIES

In the introduction to this paper it was mentioned that two bulk phases in equilibrium with each other are, in reality, separated by a transition layer in which the density of the thermodynamic properties may vary in quite a complex way from that which is characteristic of the two bulk phases. To avoid the problem of an exact description of the variation in the density of these properties across the surface layer and, thus, to be able to treat the system thermodynamically without a detailed knowledge of the structure of the transition region, Gibbs proceeds in the following manner. He selects a point in or very near the physical boundary and imagines "a geometrical surface to pass through this point and all other points which are similarly situated with respect to the condition of adjacent matter" [7]. He calls this surface SS' (Fig. 1) the *dividing surface*. The two-phase system of Fig. 1 is confined by a closed surface "such as may be generated by a moving normal to SS'" [7] extending well into the homogeneous phases on each side. The imaginary surfaces AA' and BB' are parallel to the dividing surface and lie in homogeneous regions of each phase. This system, which is characterized by a fixed energy content and total number of moles of each component (n_1, n_2, ..., n_r) is thus divided by the surface SS' into two volumes—volume V^α containing bulk phase α

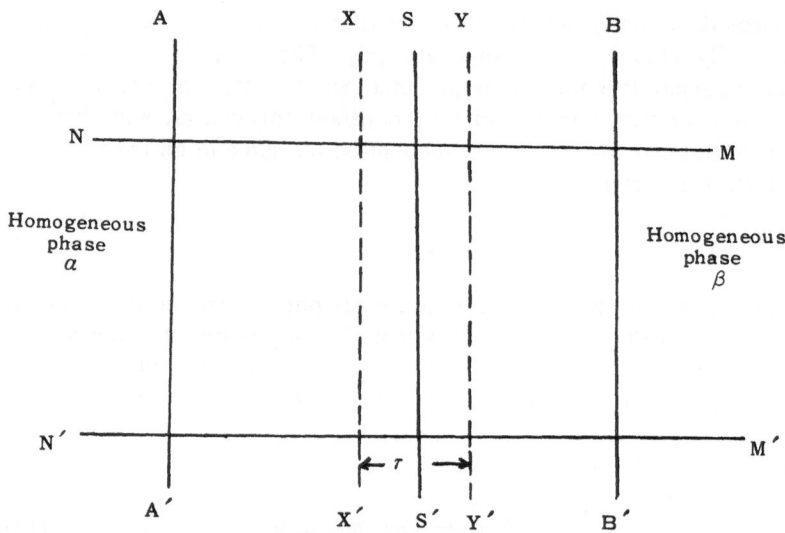

Fig. 1. Schematic presentation of a two-phase system at equilibrium. SS′—an arbitrary dividing surface. AA′, BB′—surfaces parallel to the dividing surface, but lying in the homogeneous phases α and β. XX′, YY′—surfaces defining the limit of the surface phase of Guggenheim.

together with an amount of matter in the transition zone, and volume V^β comprising the homogeneous phase β together with the remaining matter of the surface layer.

Figure 2 gives some examples of the possible variation in concentration of a component of the system in a direction normal to the dividing surface. Case A might be expected in a pure liquid–vapor system. The thickness of the interfacial region will depend on the difference in density of the two bulk phases and on the temperature. In the vicinity of the critical temperature, the surface layer should extend over large distances. The variation depicted by Case D is reminiscent of solute segregation at the grain boundary separating two identical solid phases, or it may describe the accumulation of a solute species at the interface formed between two slightly miscible phases in which the solute has a distribution coefficient of unity.

Gibbs defines the surface excess n_i^A of component i as the difference between the total number of moles of this component in a two-phase system and the number of moles in a hypothetical system of the same size in which the composition of each phase is assumed to remain

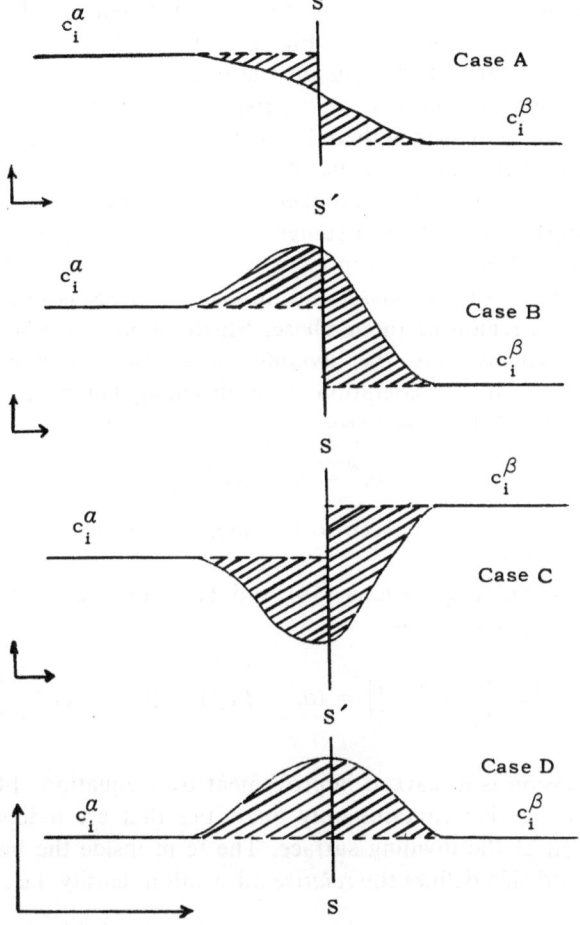

Fig. 2. Possible variations in concentration (moles/volume) of component i of a two-phase (α, β) system across the transition region.

uniform up to the dividing surface SS'. Thus, with reference to Figs. 1 and 2,

$$n_1^A \equiv A\Gamma_1 = n_1 - (c_1^\alpha V^\alpha + c_1^\beta V^\beta) \qquad (14)$$

$$n_i^A \equiv A\Gamma_i = n_i - (c_i^\alpha V^\alpha + c_i^\beta V^\beta) \qquad (15)$$

where c_i is the volume concentration of species i in the homogeneous phase; A is the surface area of the interface; and Γ_i is the surface excess per unit surface area or the adsorption density of component i.

With the exception of the variation in density depicted by Case D,* it is immediately obvious from Fig. 2 that the absolute value of the adsorption density will depend on the exact location of the dividing surface. Γ_i may be positive or negative, and a dividing surface can always be found for which there is no excess or deficit in component i. There is no restriction as to the placement of the dividing surface; it may be located within the homogeneous phase. For Case A (Fig. 2), Γ_i will vanish if SS′ is placed somewhere within the transition region; zero excess will be obtained in Case B, if SS′ is located in the homogeneous region of the β phase, and, in Case C, if SS′ is located in the homogeneous region of the α phase. Furthermore, we note that, for one-component systems, the dividing plane always can be used to eliminate Γ from consideration as a thermodynamic variable. For Case D in Fig. 2, because $c_i^\alpha = c_i^\beta$,

$$n_i^A = n_i - c_i^\alpha V \tag{16}$$

and the surface excess is seen to be independent of the placement of the dividing surface.

By combining equations (14) and (15), we may eliminate the variables V^α and V^β; thus,

$$A\left[\Gamma_i - \Gamma_1\left(\frac{c_i^\alpha - c_i^\beta}{c_1^\alpha - c_1^\beta}\right)\right] = (n_i - Vc_i^\alpha) - (n_1 - Vc_1^\alpha)\frac{c_i^\alpha - c_i^\beta}{c_1^\alpha - c_1^\beta} \tag{17}$$

This expression is a marked improvement over equation (14) because the right-hand side contains only quantities that are independent of the location of the dividing surface. The term inside the brackets on the left-hand side defines the *relative* adsorption density, i.e.,

$$\Gamma_{i(1)} \equiv \Gamma_i - \Gamma_1\left(\frac{c_i^\alpha - c_i^\beta}{c_1^\alpha - c_1^\beta}\right) \tag{18}$$

which is, by equation (17), invariant with respect to placement of the dividing surface. The experimentally meaningful expression on the right-hand side of equation (17) may also be obtained directly from equations (14) and (15) if we let $\Gamma_1 = 0$ and solve for Γ_i. The relative adsorption density $\Gamma_{i(1)}$ may, therefore, be identified as the adsorption density of the ith component at a dividing surface located so as to

---

* Another exception is the insoluble organic film at the liquid/gas interface; the organic component has a negligible concentration in both bulk phases.

make Γ_1 vanish. This choice of a dividing surface was first introduced by Gibbs [8]. With reference to equation (18), if Γ_i and Γ_1 denote the adsorption densities of the components i and 1 at an arbitrary dividing surface, then $\Gamma_1/(c_1^\alpha - c_1^\beta)$ determines the distance λ separating this dividing surface from that particular surface which makes Γ_1 vanish.

Once the dividing surface has been placed, the surface excess of all other thermodynamic quantities is defined. Thus,

$$U^A = U - (U_v^\alpha V^\alpha + U_v^\beta V^\beta) \tag{19}$$

$$\Omega^A = \Omega - (\Omega_v^\alpha V^\alpha + \Omega_v^\beta V^\beta) \tag{20}$$

$$\Omega^A = \Omega + (p^\alpha V^\alpha + p^\beta V^\beta) \tag{21}$$

and

$$U_{A(1)} \equiv U_A - \Gamma_1 \frac{U_v^\alpha - U_v^\beta}{c_1^\alpha - c_1^\beta} \tag{22}$$

$$\Omega_{A(1)} \equiv \Omega_A - \Gamma_1 \frac{p^\beta - p^\alpha}{c_1^\alpha - c_1^\beta} \tag{23}*$$

where U_v and Ω_v refer to the volume density of U and Ω in the bulk phase and $U_{A(1)}$ and $\Omega_{A(1)}$ refer to the excess per unit surface area in U and Ω, as determined by that dividing surface which makes Γ_1 vanish. The surface energy density $U_{A(1)}$ and surface free energy[†] density $\Omega_{A(1)}$ are again relative quantities that are invariant with placement of the dividing surface. The absolute value of $\Omega_{A(1)}$ is seen to be completely determined by the measurable quantities V, n_1, c_1^α, c_1^β, p^α, and p^β. In a one-component system, $U_{A(1)}$ and $\Omega_{A(1)}$ may have nonzero values, even though Γ_1 is zero.

We note again that the quantities U^A, Ω^A, S^A, etc., are not the energy, free energy, or entropy for some specific thin layer of substance between the two phases, but are correction terms (excesses) as defined. In his treatment of surface thermodynamics, Guggenheim [14] introduces a surface phase that is delineated by the two surfaces XX′ and YY′ shown in Fig. 1. This surface phase is homogeneous in directions parallel to XX′ and YY′, but varies in composition normal to these planes. The surface layer is a material system with a well-defined volume $\tilde{V}^A$, energy $\tilde{U}^A$, and free energy $\tilde{\Omega}^A$. The quantity $\tilde{n}_i^A$ is now the

---

* This expression applies to curved surfaces; for plane surfaces, $\Omega_{A(1)} = \Omega_A$.

† Throughout the text, the term free energy will refer to the Ω function; F or G will be referred to as the Helmholtz or Gibbs free energy.

actual number of moles of the ith component in the surface phase of finite thickness τ, and $\tilde{\Gamma}_i$ is the actual number of moles per unit area of surface in the surface phase. The surfaces XX′ and YY′ are to be located so that spatial variation in all thermodynamic quantities is restricted to the surface phase. We note that, although the adsorption quantity $\tilde{\Gamma}_i$ is more easily visualized physically than the quantity Γ_i referred to the imaginary Gibbsian dividing surface SS′, the former is precisely determined only if the exact location of the *two* surfaces XX′ and YY′ is known. For planar surfaces, it may be shown [15] that

$$\tilde{\Gamma}_{i(1)} \equiv \tilde{\Gamma}_i - \tilde{\Gamma}_1 \left(\frac{c_i{}^\alpha - c_i{}^\beta}{c_1{}^\alpha - c_1{}^\beta} \right)$$

$$\tilde{\Gamma}_{i(1)} = \Gamma_{i(1)} + \left(\frac{c_1{}^\alpha c_i{}^\beta - c_1{}^\beta c_i{}^\alpha}{c_1{}^\alpha - c_1{}^\beta} \right) \tau \tag{24}$$

Equations (18) and (24) assume particularly simple forms when one of the phases, say β, is a vapor of negligible density. Then, for a two-component system, because $c_1{}^\beta = c_2{}^\beta = 0$,

$$A\Gamma_{2(1)} = n_2 - n_1 \left(\frac{x^\alpha}{1 - x^\alpha} \right) \tag{25}$$

and

$$A\Gamma_{2(1)} = A\tilde{\Gamma}_{2(1)} \tag{26}$$

where

$$\tilde{\Gamma}_{2(1)} = \tilde{\Gamma}_2 - \tilde{\Gamma}_1 \left(\frac{x^\alpha}{1 - x^\alpha} \right) \tag{27}$$

Since n_2, n_1, and x^α are all positive quantities, it follows from equation (25) that in dilute solution $(x^\alpha \ll 1)$ $\Gamma_{2(1)}$ can assume large positive values, but not large negative values. When $\Gamma_{2(1)} > 0$, the solute is said to be positively adsorbed; when $\Gamma_{2(1)} < 0$, it is negatively adsorbed.

In dilute liquid solutions, the area occupied by an adsorbed molecule $\mathscr{A}$ is commonly assumed to equal the reciprocal of the relative adsorption density as follows:

$$\mathscr{A} = \frac{10^{16}}{N\Gamma_{2(1)}} \tag{28}$$

where $\mathscr{A}$ is in square angstroms. This calculation assumes that $\Gamma_{2(1)}$

represents the total amount of the solute in the surface layer—an approximation which, by equations (26) and (27), is true only if

$$\tilde{\Gamma}_2 \gg \tilde{\Gamma}_1 \left(\frac{x^\alpha}{1 - x^\alpha} \right)$$

that is, if component 2 is strongly positively adsorbed and if the solution is dilute. These restrictions have been met in the classic studies by Langmuir [16] and Harkins [17] of films of long-chained, organic, polar molecules and electrolytes formed on the surfaces of aqueous solutions.

SURFACE TENSION OR SPECIFIC SURFACE FREE ENERGY

Consider a variation in state of a two-phase multicomponent fluid system in which the initial and final states are equilibrium states. Following Gibbs, we shall regard all the properties of the surface layer as determined by the area A and the principal radii of curvature, R_1 and R_2, of some selected imaginary dividing surface. Instead of equation (9), the total differential in free energy becomes

$$d\Omega = -S \, dT - \sum n_i \, d\mu_i - p^\alpha \, dV^\alpha - p^\beta \, dV^\beta$$
$$+ \sigma \, dA + C_1 \, dR_1 + C_2 \, dR_2 \qquad (29)$$

In writing this equation, we relax the condition for mechanical equilibrium ($p^\alpha = p^\beta$) and will show later on that this condition holds only for systems with plane surfaces. The criterion for chemical equilibrium [equation (12)] still applies to systems in which surfaces are not neglected, and a rigorous proof of this is given by Gibbs [9]. The term $\sigma \, dA$ expresses the reversible work required to create dA units of surface; σ is the work required to increase the surface by one unit and is commonly referred to as the surface tension. The $C \, dR$ terms are work terms that express reversible changes in shape of the dividing surface.

From equation (29), the operational definition of surface tension is as follows:

$$\sigma = \left(\frac{\partial \Omega}{\partial A} \right)_{T,\mu,V,R} \qquad (30)$$

According to equation (30), the surface tension is measured by the change in free energy Ω of the system accompanying a unit increase in surface area at constant temperature, composition, and volume of the system and fixed curvature of the dividing surface.

Since $d\Omega$ is an exact differential, integration of equation (29), maintaining all intensive properties (including surface curvature) constant, yields

$$\Omega = \sigma A - p^{\alpha} V^{\alpha} - p^{\beta} V^{\beta} \qquad (31)$$

Since Ω is independent of the particular choice of the dividing surface, the combined terms on the right-hand side of equation (31) will also be invariant with respect to this imaginary surface. This is not necessarily true of each term taken separately and certainly not of the volumes V^{α} and V^{β}.

Combination of equations (21) and (31) yields

$$\Omega^{A} = U^{A} - TS^{A} - \sum \mu_i n_i{}^{A} \qquad (32)$$

or

$$\Omega^{A} = \sigma A \qquad (33)$$

or, per unit surface area,

$$\Omega_{A} = \sigma \qquad (34)$$

The property σ may be correctly referred to as the *specific surface free energy*, if by free energy the state function Ω is understood. The term surface tension was introduced by Gibbs with special reference to liquid surfaces. The reasons for this nomenclature will become clear when curved liquid surfaces are discussed; however, the term specific surface free energy is more appropriate for solid surfaces.

The surface tension of a system in which the two homogeneous phases are separated by a planar surface layer is an especially clearly defined and meaningful quantity. To show this, we imagine the plane dividing surface to be shifted in a direction normal to itself at constant temperature without affecting the material content of the system. For this variation,

$$\delta\Omega = 0 \qquad \delta A = 0 \qquad \delta V^{\alpha} = -\delta V^{\beta}$$

Therefore, it follows from equation (29) that $p^{\alpha} = p^{\beta}$. Thus, for plane surfaces it follows from equation (31) that

$$\sigma = \frac{\Omega}{A} + p\frac{V}{A} \qquad (35)$$

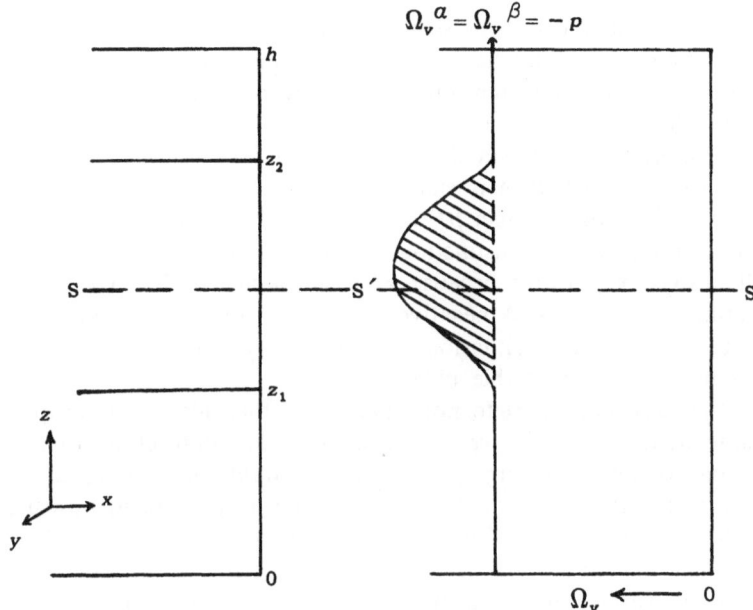

Fig. 3. Illustration of the independence of σ for plane surfaces on location of the dividing surface. The ruled area represents the excess in Ω that is equal to the surface tension σ of the plane surface.

and the surface tension is seen to be invariant with respect to the placement of the dividing surface. With reference to Fig. 3, equation (35) may be expressed in integral form as follows:

$$\sigma = \int_0^h \Omega_v(z)\, dz + \int_0^h p\, dz \tag{36}$$

where h denotes the extent of the system in the z direction normal to the dividing surface. The independence of σ on the location of a plane dividing surface as expressed by equation (36) is illustrated in Fig. 3, which should be compared with Case D of Fig. 2.

Equation (36) may be replaced by the following relation:

$$\sigma = \int_0^h (p - p')\, dz \tag{37}$$

which is an expression similar in form to that obtained by the continuum mechanical treatment of capillarity by Bakker, where the term $(p - p')$

denotes the difference in pressure normal and tangential to any thin film of fluid parallel to the surface. According to Bakker's treatment, the existence of surface tension is an expression of the deviation from Pascal's law in the surface layer.

Usually, σ is very small, e.g., of the order of 10–1000 ergs/cm² (10^{-6}–10^{-4} cal/cm²). If A is small, then Ω will be almost exclusively determined by the product pV in equation (35). It is also important to note that, in contradistinction to adsorption densities, σ and Ω_A will always be positive because work must always be expended in creating new surfaces. A negative surface tension indicates instability of the surface; such a situation must inevitably lead to the formation of a new and more stable phase.

It is also important to note that σ is a thermodynamic property characteristic of solid as well as liquid surfaces. Gibbs clearly pointed out that for solids σ (energy per unit area) should not be confused with surface stress (force per unit length), the latter property being measured by the reversible work of deforming a surface and the former being determined by the reversible work of creating a new surface.

If the surface area of a pure liquid is increased by an isothermal deformation at constant volume, then the change in free energy of the liquid is given by the following expression:

$$d\Omega = \sigma \, dA + A \, d\sigma$$
$$d\Omega = \gamma \, dA \tag{38}$$

where γ is the surface stress (force per unit length). Equation (38) may be transformed to give the following relation between surface stress and σ:

$$\gamma = \sigma + A \frac{d\sigma}{dA} \tag{39}$$

The second term on the right-hand side of equation (39) will be zero because, in this deformation, liquid molecules will be transferred to the surface and, thus, the number of molecules per unit area in the surface layer will remain unaltered. Therefore, we conclude that, for liquids, surface stress γ is numerically equal to σ. This equality of σ and γ also follows because liquids are unable to support shear stresses and will relax completely when the forces responsible for the deformation are released. In fact, most of the experimental methods for determining the surface tension σ of liquids involve balancing the surface stress with a measurable counterforce. From an atomistic viewpoint, the surface stress γ results from the net inward pull of the molecules in the bulk liquid on the surface molecules, which are not symmetrically surrounded by like neighbors. The surface stress is visualized as a tension in the surface; thus, the term surface tension for σ.

Surface stress in solids has the same atomistic origin as the tensile surface stress in liquids. It is related to the rearrangements and configurational changes experienced by the atoms in the surface layer in an attempt to minimize their potential energy. A distortion of the surface layers relative to similar planes in the bulk crystal will result in and will manifest itself as a compressive or tensile surface stress, if the interaction with atoms in nonadjacent planes is not negligible. If they are not relieved by the application of external forces, these surface stresses will be balanced by nonuniform volume stresses induced in the bulk solid. In small crystals ($< 0.1\ \mu$ in size), the existence of these surface stresses and associated volume stresses cannot be completely ignored in the thermodynamic description of the solid. The distinction between surface stress and surface free energy σ for solids may be made clear by postulating the existence of a homopolar solid in which only nearest neighbor atoms interact. For this hypothetical solid, there will be no surface distortion and, therefore, no surface stress; the atom spacing in the surface layer will be identical to that in similar planes in the bulk. Nevertheless, the surface will have a characteristic nonzero σ. Furthermore, σ may be expected to vary with crystallographic orientation of the surface plane because of the varying arrangements of atoms near different surfaces.

A straightforward application of equation (39) to solids is not possible but it is obvious that $d\sigma/dA$ for a given surface plane will be zero only if this is a high-symmetry plane and if the deformation results in an infinitesimal change of shape, but not in surface area. At relatively low temperatures, the atoms are not mobile. Thus, after any deformation,* they maintain their relative positions, even though the separation between the atoms will be changed; also, the total number of atoms in the surface remains unchanged. Clearly, σ is then a function of A, and $d\sigma/dA$ cannot be zero [25].

It is not the purpose of this paper to give a detailed description of the surface properties of solids. The reader is referred for this to

---

* The initial state before deformation must be one of zero volume strain in order to apply equation (39). See Shuttleworth [25].

the excellent treatments by Herring [18], Mullins [19], and Cabrera [20]. In conclusion, we recall that the equilibrium form of a homogeneous body (solid or liquid) is determined by the condition that Ω shall be a minimum for given values of temperature, chemical potential, and volume V of the body. For crystals, this condition is expressed by the Gibbs–Curie criterion as follows:

$$\delta(\sum \sigma_i A_i) = 0$$

$$\delta^2(\sum \sigma_i A_i) > 0 \tag{40}$$

subject to the constraints of constant temperature and volume of the crystal. In equation (40) σ_i refers to the surface free energy of the ith crystal face with surface area A_i. For isotropic liquids, the equilibrium form is determined by the condition that its total surface should be a minimum, that is, it becomes a sphere. Deviations from spherical shape will occur when the effect of gravity is taken into consideration, but with decreasing size the departure from sphericity becomes less. In homogeneous nucleation systems, the critical liquid nucleus may always be assumed to have a spherical shape.

THE GIBBS ADSORPTION EQUATION FOR PLANE SURFACES

On subtracting the contribution of bulk phases [equation (9)] to the total differential $d\Omega$ in equation (29), it follows that for a plane surface

$$d(\Omega - \Omega^\alpha - \Omega^\beta) = d\Omega^A = -S^A \, dT - \sum n_i{}^A \, d\mu_i + \sigma \, dA \tag{41}$$

Differentiation of equation (33) and comparison of the result with equation (41) shows that

$$A \, d\sigma = -S^A \, dT - \sum n_i{}^A \, d\mu_i \tag{42}$$

and, because σ is independent of the extent of the surface,*

$$d\sigma = -S_A \, dT - \sum^r \Gamma_i \, d\mu_i \tag{43}$$

This expression, which relates changes in surface tension to variations in temperature and chemical potential, was first derived by Gibbs and

* It is assumed that for solids the orientation of the exposed surface is not changed during the variation.

is known as the Gibbs adsorption equation. It is recognized as the surface equivalent of the Gibbs–Duhem equation for homogeneous phases. The condition for thermodynamic equilibrium expressed by equation (12) still applies, and, in addition, the equivalence of the chemical potential of a component in the surface with its value in a homogeneous phase is required. It should be noted that this equilibrium criterion is strictly valid only if the action of gravitational or other force fields can be neglected; if not, it is the sum of the chemical potential and the molal potential energy of the component in the external field, which remains uniform throughout the system.

Application of the phase rule to surface systems shows that σ will be completely described by r of the $r + 1$ variables appearing on the right-hand side of equation (43). The chemical potential of one of the components of the system, e.g., component 1, may thus be eliminated from equation (43) with the aid of equation (13). Thus, at constant temperature,

$$d\sigma = -\sum_{2}^{r} \left[\Gamma_i - \Gamma_1 \left(\frac{c_i^\alpha - c_i^\beta}{c_1^\alpha - c_1^\beta} \right) \right] d\mu_i \qquad (44)$$

and, in view of equation (18),

$$d\sigma = -\sum_{2}^{r} \Gamma_{i(1)} \, d\mu_i \qquad (45)$$

This is the form in which the Gibbs equation is applied to experimental adsorption studies. We note that the very peculiar dividing surface introduced by Gibbs leads, nevertheless, to quantities which are experimentally meaningful.

We digress briefly to point out that the Gibbs adsorption equation [equation (45)] may be derived without introducing a dividing surface. For plane–surface systems, the Gibbs–Duhem equation becomes

$$A \, d\sigma - V \, dp + \sum n_i \, d\mu_i + S \, dT = 0 \qquad (46)$$

Multiplication of the Gibbs–Duhem expression for the homogeneous phase α

$$dp - S_v^\alpha \, dT - \sum c_i^\alpha \, d\mu_i = 0 \qquad (47)$$

by the total volume V of the system and use of the result to eliminate $V \, dp$ from equation (46) yields

$$A \, d\sigma = -\sum (n_i - V c_i^\alpha) \, d\mu_i - (S - V S_v^\alpha) \, dT \qquad (48)$$

Further elimination of the chemical potential μ_1 from equation (48) at constant temperature shows that, without ever introducing the concept of a dividing surface,

$$d\sigma = -\sum_{2}^{r} \left[(n_i - Vc_i^{\alpha}) - (n_1 - Vc_1^{\alpha}) \left(\frac{c_i^{\alpha} - c_i^{\beta}}{c_1^{\alpha} - c_1^{\beta}} \right) \right] d\mu_i / A \qquad (49)$$

which is the exact equivalent of equation (45).

Equation (45) shows that in a binary system positive adsorption of component 2 ($\Gamma_{2(1)} > 0$) leads to a lowering of the surface tension, and negative adsorption ($\Gamma_{2(1)} < 0$) results in an increase in surface tension. Furthermore, in view of the previously presented general discussion of adsorption, positive adsorption could lead to a drastic lowering in surface tension in dilute solutions of a surface-active species, but negative adsorption in dilute solution can give rise to only a slight increase in surface tension. These deductions have been confirmed by surface-tension measurements of numerous dilute aqueous solutions.

For a binary system, the dependence of surface tension on composition x at constant temperature and pressure* is, from equation (45), as follows:

$$\begin{aligned} \frac{d\sigma}{dx} &= -\Gamma_{2(1)} \frac{d\mu_2}{dx} \\ \frac{d\sigma}{dx} &= -RT\Gamma_{2(1)} \frac{d \ln a_2}{dx} \end{aligned} \qquad (50)$$

From measurements of σ and μ (or activity) over a wide range of values of x (mole fraction of component 2 in, for example, phase α), the adsorption density $\Gamma_{2(1)}$ may be calculated as a function of x. Depending on the exact nature of the variation of σ and μ_2 with x, $\Gamma_{2(1)}$ when positive may pass through a maximum. The condition for maximum (or minimum) in $\Gamma_{2(1)}$ to be present is expressed by the following identity:

$$\frac{d^2\sigma}{dx^2} \left(\frac{d\mu_2}{dx} \right) = \frac{d^2\mu_2}{dx^2} \left(\frac{d\sigma}{dx} \right) \qquad (51)$$

The various types of σ–x curves observed in binary systems are sketched in Fig. 4; superimposed on this plot is the variation of μ_2

---

* For liquid–vapor systems, the pressure is not an independent variable, but is a function of the composition of the liquid phase at constant temperature.

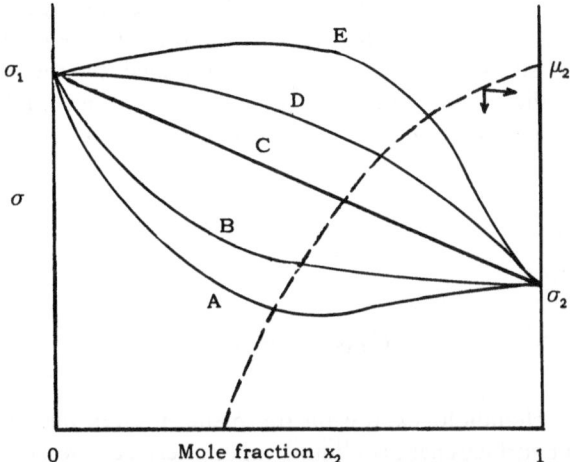

Fig. 4. Possible variation of σ with composition in binary systems. The broken line represents the dependence of chemical potential of component 2 on composition.

with x. Application of the identity [equation (51)] to these curves shows that $\Gamma_{2(1)}$ may pass through a maximum if Curve B represents the variation of σ with x for a given system; and Curves C and D will give no maximum or minimum in $\Gamma_{2(1)}$. Curve A must lead to a maximum in $\Gamma_{2(1)}$, and Curve E, to a minimum in $\Gamma_{2(1)}$; in both instances, $\Gamma_{2(1)}$ becomes zero at some value of x intermediate between zero and unity.

APPLICATION

Single-Component Systems

According to the first two laws of thermodynamics, the increase in internal energy due to the reversible formation of one unit of surface at constant volume and temperature is given by the following relation:

$$\left(\frac{\partial U}{\partial A}\right)_{T,V} = T\left(\frac{\partial S}{\partial A}\right)_{T,V} + \sigma \tag{52}$$

Since the adsorption density can always be made to vanish by proper placement of the dividing surface, it also follows from equations (22) and (33) that

$$U_{A(1)} = TS_{A(1)} + \sigma \tag{53}$$

Comparison of these two expressions shows that the specific surface energy $U_{A(1)}$ is determined by the increase in internal energy and that the specific surface entropy $S_{A(1)}$ is determined by the increase in entropy of the system due to a reversible isothermal and isochoric creation of one unit of surface. Furthermore, for one-component systems,

$$do = -S_{A(1)} \, dT \tag{54}$$

and, therefore,

$$U_{A(1)} = \sigma - T \frac{d\sigma}{dT} \tag{55}$$

This Gibbs–Helmholtz expression for plane surfaces may be used to calculate the surface energy of liquid/vapor interfaces from experimental measurements of surface tension as a function of temperature. The temperature coefficient of the surface tension of pure liquids will be negative because σ must obviously vanish at the critical temperature.

From a molecular viewpoint, the negative temperature coefficient of surface tension may be explained by the decreasing attractive forces between adjacent liquid molecules due to an increase in their average separation with rising temperature. The increased concentration of vapor molecules with increasing temperature also contributes to this decrease in σ by opposing more strongly the inward attraction of bulk liquid molecules on the surface molecules. At temperatures far below the critical temperature, the specific surface energy usually varies only slightly with temperature, whereas σ quite often varies linearly with temperature. Mercury at ordinary temperatures is a good example of a liquid with constant specific surface energy, since it has a low vapor pressure and a comparatively small volume expansion coefficient.

Multicomponent Systems

For liquid–vapor systems, temperature and composition of the liquid phase are logical choices as independent variables. The composition variables may be either the mole fraction x of the components in the liquid phase or, if the solute components are volatile, the partial pressures in the saturated vapor phase. In very dilute solutions, the decrease in surface tension is, as a first approximation, proportional to the mole fraction (or molar concentration) of the positively adsorbed solute species. From the Gibbs adsorption equation, it may be seen that if Henry's law [equation (6)] holds, $\Gamma_{i(1)}$ is also proportional to the concentration of the solute species.

An interesting example of the two-component system is the solid–solid system in which two identical phases are separated by a grain boundary where equilibrium segregation (positive adsorption) of a solute species may be observed. At constant temperature, the effect of changes in solute concentration in the bulk phase on the grain boundary tension σ is given by the following expression, which was obtained by combination of equations (11b) and (43):

$$d\sigma = -\left(\Gamma_2 - \Gamma_1 \frac{x}{1-x}\right) d\mu_2 \quad (56)$$

where the mole fraction x refers to the solute species in both bulk phases. With reference to equation (16), the quantity in parentheses may be equated to $\Gamma_{2(1)}$ and is simply related to experimental quantities as follows:

$$\Gamma_{2(1)} = \frac{1}{A}\left(n_2 - n_1 \frac{x}{1-x}\right) \quad (57)$$

Cahn and Hilliard [21] showed how the Gibbs adsorption equation may be used to determine whether the observed excess of a solute component at the grain boundary is present as an equilibrium segregate or as a precipitate (new phase).

In contrast to liquid–vapor systems, it is convenient to choose both pressure and temperature as independent variables in those systems where the interface is formed between two condensed phases (liquid–liquid, liquid–solid, and solid–solid). The Gibbs equation [equation (45)] may be applied to these systems, but we may also wish to retain pressure as an independent variable. This means that, for a two-phase system of r components, changes in surface tension must be expressed as functions of T, p, and $(r - 2)$ chemical potentials. As was already pointed out by Gibbs [10], to accomplish this it is necessary to introduce two dividing surfaces such that the placement of one causes the adsorption density of, e.g., component 1 to vanish; and the placement of the other dividing surface makes the adsorption density of the other component, e.g., 2, vanish. The following analysis was first performed by Hansen [22].

If the following fundamental relations:

$$A \, d\sigma = -S \, dT + V \, dp - \sum n_i \, d\mu_i \tag{58}$$

$$dp = S_v{}^\alpha \, dT + \sum c_i{}^\alpha \, d\mu_i \tag{59}$$

$$dp = S_v{}^\beta \, dT + \sum c_i{}^\beta \, d\mu_i \tag{60}$$

that are characteristic of the total system and of each homogeneous phase are used, the desired adsorption equation is obtained by eliminating μ_1 and μ_2 from equation (58) by introducing the Lagrangian multipliers λ^α and λ^β. Multiplication of equations (59) and (60) by λ^α and λ^β, respectively, and substraction of these results from equation (58) yields

$$A d\sigma = -(S - \lambda^\alpha S_v{}^\alpha - \lambda^\beta S_v{}^\beta) \, dT + (V - \lambda^\alpha - \lambda^\beta) \, dp$$

$$- \sum (n_i - \lambda^\alpha c_i{}^\alpha - \lambda^\beta c_i{}^\beta) \, d\mu_i \tag{61}$$

Two differentials in equation (61) may be eliminated by choosing λ^α and λ^β such that their coefficients are zero. We note that λ has the dimension of volume and λ/A that of length. If the following substitutions are used,

$$S - \lambda^\alpha S_v{}^\alpha - \lambda^\beta S_v{}^\beta \equiv AS_A \tag{62}$$

$$V - \lambda^\alpha - \lambda^\beta \equiv A\tau \tag{63}$$

and

$$n_i - \lambda^\alpha c_i{}^\alpha - \lambda^\beta c_i{}^\beta \equiv A\Gamma_i \tag{64}$$

equation (61) reduces to

$$d\sigma = -S_A \, dT + \tau dp - \sum \Gamma_i \, d\mu_i \tag{65}$$

If λ^α and λ^β are chosen such that $\Gamma_1 = \Gamma_2 = 0$, then

$$d\sigma = -S_{A(1,2)} \, dT + \tau_{(1,2)} \, dp - \sum_3^r \Gamma_{i(1,2)} \, d\mu_i \tag{66}$$

which is the desired relationship. If λ^α and λ^β are chosen such that $\tau = 0$ and $\Gamma_1 = 0$, the Gibbs adsorption equation results:

$$d\sigma = -S_{A(1)} \, dT - \sum_2^r \Gamma_{i(1,2)} \, d\mu_i \tag{67}$$

Note that if $\Gamma_1 = \Gamma_2 = 0$, then

$$-\Gamma_{i(1,2)} = \left(\frac{\partial\sigma}{\partial\mu_i}\right)_{T,p,\mu_3,\ldots,\mu_{i-1},\mu_{i+1},\ldots,u_r} \tag{68}$$

whereas

$$-\Gamma_{i(1)} = \left(\frac{\partial\sigma}{\partial\mu_i}\right)_{T,\mu_2,\ldots,\mu_{i-1},\mu_{i+1},\ldots,u_r} \tag{69}$$

and

$$\lambda^\alpha = \frac{n_1c_2{}^\beta - n_2c_1{}^\beta}{c_1{}^\alpha c_2{}^\beta - c_1{}^\beta c_2{}^\alpha} \tag{70}$$

$$\lambda^\beta = \frac{n_2c_1{}^\alpha - n_1c_2{}^\alpha}{c_1{}^\alpha c_2{}^\beta - c_1{}^\beta c_2{}^\alpha} \tag{71}$$

The application of this analysis to a two-component system shows that the adsorption equation reduces to

$$d\sigma = -S_{A(1,2)}\,dT + \tau_{(1,2)}\,dp \tag{72}$$

and that

$$U_{A(1,2)} = TS_{A(1,2)} - \tau_{(1,2)}p + \sigma$$
$$U_{A(1,2)} = \sigma - T\left(\frac{\partial\sigma}{\partial T}\right)_p + p\left(\frac{\partial\sigma}{\partial p}\right)_T \tag{73}$$

or

$$H_{A(1,2)} = \sigma - T\left(\frac{\partial\sigma}{\partial T}\right)_p \tag{74}$$

where

$$H_{A(1,2)} = \left(\frac{\partial H}{\partial A}\right)_{p,T} \tag{75}$$

and

$$U_{A(1,2)} = \left(\frac{\partial U}{\partial A}\right)_{p,T} \tag{76}$$

Equation (74) should be compared with equation (55), which applies to one-component systems.

For a three-component system at constant temperature and pressure,

$$d\sigma = -\Gamma_{3(1,2)}\,d\mu_3 \tag{77}$$

where

$$\Gamma_{3(1,2)} = \frac{1}{A}\left[n_3 + n_2\left(\frac{c_3{}^\alpha c_1{}^\beta - c_3{}^\beta c_1{}^\alpha}{c_1{}^\alpha c_2{}^\beta - c_1{}^\beta c_2{}^\alpha}\right) + n_1\left(\frac{c_3{}^\beta c_2{}^\alpha - c_3{}^\alpha c_2{}^\beta}{c_1{}^\alpha c_2{}^\beta - c_1{}^\beta c_2{}^\alpha}\right)\right] \quad (78)$$

The relative adsorption density $\Gamma_{3(1,2)}$ may, therefore, be calculated from experimental quantities. The adsorption equation [equation (77)] may be compared with the expression

$$d\sigma = -\Gamma_{2(1)}\,d\mu_2 - \Gamma_{3(1)}\,d\mu_3 \quad (79)$$

where pressure is not an independent variable.

Surface Films

The resistance of two-component foam systems to collapse by local deformation of the liquid lamellae can be understood by combining the Gibbs adsorption equation with equation (39). If we consider an extension of the liquid lamellae at constant temperature and total number of moles of the surface-active solute and the solvent, then the term $A(d\sigma/dA)$ will not be zero. An increase in surface area will result in an increase in σ, because the Gibbs adsorption equation predicts that a decrease in the chemical potential of the solute in the solution must lead to an increase in σ if the solute is surface-active. This increase in σ tends to resist any further extension of the lamellae and, therefore, helps to prevent collapse of the foam by surface distortion. Gibbs [13] defined the modulus of surface elasticity

$$E \equiv 2A\,\frac{d\sigma}{dA} \quad (80)$$

and showed that it is related to the relative adsorption density of the solute (component 2) as follows:

$$E = 4\Gamma_{2(1)}^2\left(\frac{\partial\mu_2}{\partial\Phi_2}\right)_{\Phi_1} \quad (81)$$

where Φ_1 and Φ_2 are the total quantities of the components per unit area of film. Equation (81) implies that the deformation is performed at an infinitely small rate, because equilibrium between surface and bulk must be maintained in order that the Gibbs adsorption equation may be applied. However, deformations in foam lamellae may be so rapid that equilibrium may be disturbed for long periods depending on the diffusion rate of the chemical species. The surface tension increase following a rapid surface extension may, therefore, be larger than that predicted by equation (81).

CURVED SURFACES

This discussion of curved surfaces applies to liquid–liquid and liquid–vapor systems. As shown previously, the expression

$$\sigma = \frac{1}{A}(\Omega + p^\alpha V^\alpha + p^\beta V^\beta) \tag{82}$$

applies to any form or shape of the dividing surface, but only in the case of plane surfaces will σ be invariant with respect to the location of this mathematical surface. According to equations (29) and (41), for a curved boundary,

$$(d\Omega^A)_{T,\mu} = \sigma\, dA + C_1\, dR_1 + C_2\, dR_2 \tag{83}$$

Differentiation of equation (33) and comparison with equation (83) yields

$$A\, d\sigma = C_1\, dR_1 + C_2\, dR_2 \tag{84}$$

where T and μ are constant. If we now assume a spherical dividing surface and imagine an isothermal displacement of this surface normal to itself while keeping the physical content of the system unaltered, then

$$\delta\Omega = 0 \qquad \delta A = 2A\,\delta R/R \qquad \delta V^\alpha = A\,\delta R = -\delta V^\beta \tag{85}$$

and

$$\delta(\sigma A) = (p^\alpha - p^\beta)\,\delta V^\alpha \tag{86}$$

or, after substitution for δV^α and δA,

$$A\,\delta\sigma + \left(\frac{2\sigma}{R}\right) A\,\delta R = (p^\alpha - p^\beta) A\,\delta R \tag{87}$$

The important relation

$$\left(\frac{\partial\sigma}{\partial R}\right)_{T,\mu} + \frac{2\sigma}{R} = p^\alpha - p^\beta \tag{88}$$

follows from equation (87). It reduces to the familiar Laplace equation of capillarity

$$\frac{2\sigma_m}{R_m} = p^\alpha - p^\beta \tag{89}$$

where σ_m and R_m are quantities describing that spherical dividing surface at which the condition

$$\left(\frac{\partial \sigma}{\partial R}\right)_{T,\mu} = 0 \tag{90}$$

holds. Gibbs calls this particular dividing surface the *surface of tension*, because the Laplace equation implies that the mechanical effect of a curved interfacial region of complicated structure may be compared to a stretched membrane of zero thickness separating two homogeneous fluids at pressures p^α and p^β. This fictitious membrane is characterized by a uniform tension σ_m in all directions; this is why Gibbs elected to call this thermodynamic property *surface tension*. As mentioned in a previous section, confusion may arise if, by the surface tension of solid surfaces, the existence of a tensile stress is implied. The term specific free energy ($\sigma = \Omega_A$) will avoid such misinterpretation.

Comparison of equations (84) and (90) shows that, at the surface of tension for spherical boundaries,

$$\left(\frac{\partial \sigma}{\partial R}\right)_{T,\mu} = (C_1 + C_2) \equiv C = 0 \tag{91}$$

At the surface of tension, therefore, the coefficient $(C_1 + C_2)$* vanishes, as noted by Gibbs. For spherical surfaces, the difference $(C_1 - C_2)$ is always zero, a condition which is not met with nonspherical curvatures.

According to equation (90), σ_m may be a maximum or a minimum in σ. To show that σ_m is a minimum for a system of fixed temperature, composition, and volume V, we add the product $p^\beta V$ to both sides of equation (82). After rearranging terms, we find

$$\sigma A - (p^\alpha - p^\beta)V^\alpha = \Omega + p^\beta V = \text{constant} \tag{92}$$

If a spherical boundary to separate phase α from the surrounding phase β is assumed, the constant in equation (92) may be evaluated with the aid of the Laplace equation [equation (89)]. Thus,

$$\sigma R^2 - \frac{2}{3}\left(\frac{\sigma_m}{R_m}\right) R^3 = \frac{1}{3}\sigma_m R_m^2 \tag{93}$$

---

* For nonspherical surfaces, equation (84) according to Gibbs may be written in the following form:
$$A d\sigma = \frac{1}{2}(C_1 + C_2) d(R_1 + R_2) + \frac{1}{2}(C_1 - C_2) d(R_1 - R_2)$$
and, at the surface of tension, the term $\frac{1}{2}(C_1 + C_2)$ vanishes.

or, on rearranging terms,

$$\frac{\sigma}{\sigma_m} = \tfrac{2}{3}\frac{R}{R_m} + \tfrac{1}{3}\frac{R_m{}^2}{R^2} \tag{94}$$

where R is the radius of an arbitrary spherical dividing surface. Equation (94) clearly shows that σ_m is a minimum and that the minimum is unique. A graphical presentation of this expression is given in Fig. 5. We note that if R_m is large compared to the thickness of the interfacial layer, then all those σ's relating to dividing surfaces located within or near the surface layer, within which the surface of tension lies, are approximately equal to σ_m. As Gibbs pointed out, this means that from a macroscopic standpoint we may regard the surface tension of curved liquid surfaces as independent of the location of the dividing surface, provided it is placed within the surface layer. If we choose the dividing surface to make Γ_i of a solute vanish, then it is possible that this condition (equality of σ with σ_m) may not be satisfied. See, for example, Fig. 2 and the accompanying discussion.

The condition for mechanical equilibrium in a system in which two homogeneous phases are separated by a surface of spherical curvature is expressed by the Laplace equation [equation (89)]. The radius of curvature R_m is to be taken as positive when its center lies

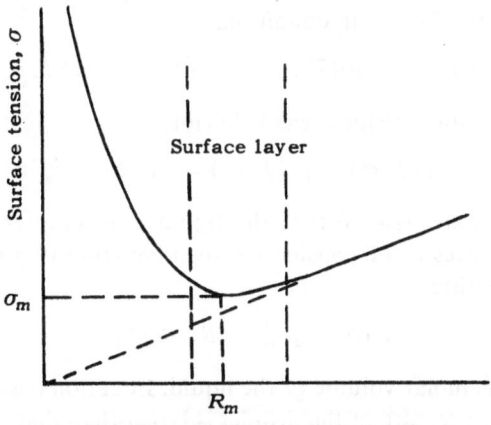

Fig. 5. Variation of surface tension with radius of spherical dividing surface at constant T and μ.

in the phase to which p^α relates. In general, for any curved surface, the harmonic mean radius of curvature

$$\frac{1}{R} = \tfrac{1}{2}\left(\frac{1}{R_1} + \frac{1}{R_2}\right) \tag{95}$$

must replace the spherical radius of curvature R_m in equation (89). The Laplace equation forms the basis of the capillary rise (or depression), the maximum bubble pressure, and all other techniques for measuring the surface tension of liquids. For a spherical water droplet ($\sigma = 72$ dynes/cm at 25°C) of radius 10^{-5} cm, the pressure inside the liquid exceeds that in the vapor phase surrounding it by 14.2 atm. Fantastically large pressure differences are indicated for very small droplets; however, one would expect the surface tension itself to become a function of the curvature if the droplet radius is of the order of the thickness of the interfacial layer. For curved surfaces, the condition for chemical equilibrium still requires the equality of the chemical potential of an actual component throughout the system. It is obvious, however, that because the pressures of the homogeneous phases are no longer equal, the chemical potential of a liquid droplet must exceed that of an infinitely large drop ($R = \infty$) at the same temperature. The difference in chemical potential of a liquid droplet at pressure p^α and the same liquid (phase α) when separated by a plane surface from its saturated vapor pressure p_∞ at temperature T is expressed by the following condition:

$$\mu^\alpha(T, p^\alpha) - \mu^\alpha(T, p_\infty) = \mu^\beta(T, p^\beta) - \mu^\beta(T, p_\infty) \tag{96}$$

and, on assumption of ideal gas behavior,

$$\mu^\alpha(T, p^\alpha) - \mu^\alpha(T, p_\infty) = RT \ln p^\beta/p_\infty \tag{97}$$

If we furthermore assume that the liquid is incompressible over the range of pressures to be considered, then equation (97) reduces to the following identity:

$$v^\alpha(p^\alpha - p_\infty) = RT \ln p^\beta/p_\infty \tag{98}$$

where v^α is the molar volume of the liquid. Equation (98) shows clearly that the vapor pressure of the droplet is larger than that of an infinitely large mass of the same liquid. Experiment shows that the increase in vapor pressure ($p^\beta - p_\infty$) is always much smaller than the pressure drop ($p^\alpha - p^\beta$) across the curved surface; thus, p_∞ may be replaced by the vapor pressure of the droplet in equation (98) and, consequently, by equation (89),

$$RT \ln p^\beta/p_\infty = \frac{2v^\alpha \sigma_m}{R_m}$$ (99)

This expression is the familar Kelvin equation,* which relates changes in vapor pressure to droplet size as measured by the radius of the spherical surface of tension. In this equation, the molar volume may be replaced by the ratio M/ρ^α, where M is the molecular weight and ρ^α is the density of the pure liquid. For a liquid solution, p^β and p_∞ will refer to the partial pressures of a volatile component, and v^α should be replaced by the partial molal volume of the same component in the solution phase.

If the surface tension (or specific free energy) of a solid is independent of crystallographic orientation, the equilibrium shape of the crystal will be that of a sphere, and equation (99) may also be applied to this system. An analogous expression relating the solubility of the crystal in a solvent as a function of its size has been used to determine the surface tension of the solid/liquid interface. Normally, however, the surface tension σ of a crystal is a function of crystallographic orientation of the surface, and the dependence of vapor pressure on crystal size will be expressed as follows:

$$RT \ln p^\beta/p_\infty = \frac{2v^\alpha \sigma_i}{l_i}$$ (100)

where l_i is the length of a perpendicular dropped from the geometric center of the crystal to the face A_i of surface tension σ_i if the equilibrium shape can be described by a regular polyhedron. Equation (100) may also be used to determine the equilibrium shape of a crystal;

$$\frac{\sigma_1}{l_1} = \frac{\sigma_2}{l_2} = \cdots = \frac{\sigma_i}{l_i} = \frac{RT}{2v^\alpha} \ln \frac{p^\beta}{p_\infty} = K$$ (101)

The equilibrium shape will include only those crystal faces for which the ratio σ_i/l_i is equal to the Wulff constant K. Herring [23] showed that the equilibrium shape need not necessarily be a regular polyhedron; it may also be bounded partly or completely by smoothly curved surfaces.

If saturated water vapor at 25°C ($p_\infty = 23.8$ mm Hg) is compressed to 26.4 mm Hg at constant temperature, according to equation (100),

---

* Also referred to as the Gibbs–Thomson equation.

liquid droplets of radii 10^{-6} cm would just be in equilibrium with the compressed vapor; smaller drops, if present, will evaporate and larger drops will grow in size. Stable equilibrium will be reached by coalescence of the metastable drops of radius 10^{-6} cm and by continued growth of larger drops until the vapor pressure has dropped to 23.8 mm Hg if the temperature is maintained constant. Similarly, with reference to equation (101), those crystal faces for which $\sigma_i/l_i \neq K$ will either evaporate or dissolve if $\sigma_i/l_i > K$ or else grow out of existence if $\sigma_i/l_i < K$. These remarks are mainly applicable to small crystals; kinetic effects are more important in controlling the shape of large crystals [18,23]. For spherical surfaces, in view of equations (84) and (88), the general form of the Gibbs adsorption equation states that at constant temperature

$$do = -\sum \Gamma_i \, d\mu_i + \left(\frac{\partial \sigma}{\partial R}\right)_{T,\mu} dR \qquad (102)$$

Only for the surface of tension is the expression

$$d\sigma_m = -\sum \Gamma_{i(m)} \, d\mu_i \qquad (103)$$

found which resembles the adsorption equation for plane surfaces. If we choose for the dividing surface a location which would make Γ_1 vanish and represent the surface tension of this spherical surface by σ_v and its radius of curvature by R_v, then instead of equation (103) the following expression applies:

$$d\sigma_v = -\sum_2^r \Gamma_{i(1)} \, d\mu_i + \left(\frac{\partial \sigma}{\partial R}\right)_{T,\mu,R=R_v} dR \qquad (104)$$

If the radius of curvature is large compared to the thickness of the surface layer, the variation of σ with radius of curvature at constant T and μ may be neglected, and the Gibbs adsorption equation for plane surfaces [equation (45)] may also be applied to curved boundaries.

For one-component systems, it follows from equations (102) and (104) that

$$\left(\frac{\partial \sigma_v}{\partial R_v}\right)_T = \left(\frac{\partial \sigma}{\partial R}\right)_{T,\mu,R=R_v} \qquad (105)$$

This relationship may be used to evaluate the dependence of surface tension on the size of a liquid droplet [24].

THERMODYNAMICS OF HOMOGENEOUS NUCLEATION

We will consider briefly the role of surface thermodynamics in evaluating the minimum work required to form a critical nucleus. In a homogeneous phase, e.g., a vapor, because of fluctuations, there will appear small aggregates of another phase, e.g., liquid droplets. If the vapor is the stable phase, then these globules will be unstable and will disappear with time. However, if the vapor is supercooled and thereby supersaturated, droplets of sufficient size will be in unstable equilibrium with the metastable vapor phase and will tend to grow in time through collisions with vapor molecules. The finite dimensions of the droplets are required to balance the loss of energy accompanying the creation of an interface between the vapor and liquid phases. Thus, for any metastable phase, there exists some minimum dimension which a particle of another phase, the critical nucleus, must have if it is to be more stable than the parent phase.

The equilibrium between the critical nucleus and the continuous phase (e.g., the supercooled vapor) is described by the following requirement:

$$\mu_o(T, p_o, x_o) = \mu_n(T, p_n, x_n) \tag{106}$$

where the subscript o refers to the parent (outside) phase and the subscript n to the critical nucleus.

If the nucleus is so small that no part of it can be homogeneous, equation (106) may still be applied if, as suggested by Gibbs [11], we imagine the space occupied by the nucleus material to be filled with bulk matter at that pressure and composition at which this assumed homogeneous phase will be in equilibrium with the exterior phase. Equation (106) expresses the condition of equilibrium between any two phases of a chemical substance ($x = 1$) separated by a semipermeable barrier which prevents the transmission of pressure, but which does not interfere with the transfer of matter between the two bulk phases. The pressure p_n is given by the Gibbs–Poynting relation [equation (98)], but this will not be the pressure at the center of the droplet if the nucleus is nonuniform throughout. Experiment has shown that a measurable rate of condensation of a liquid from a supersaturated vapor phase is normally obtained at a degree of supersaturation (p/p_∞) which, according to equation (99), predicts the presence of nonuniform critical nuclei.

Next we consider the formation of the critical nucleus in a very large volume V of the homogeneous parent phase (at pressure p_o) at constant temperature, composition, and volume. The reversible work of nucleation $W*$ is then equal to the change in free energy of the system—

$$W* = (\Delta\Omega)_{T,\mu,V}$$
$$W* = \Omega_{II} + p_o V \tag{107}$$

where Ω_{II} is the free energy of the system, including the nucleus of the same total volume V as the initial completely homogeneous system. We note that the expression on the right-hand side of equation (107) also occurs in equation (92), and, thus, physical reality is given to the purely mathematical operation by which the latter equation was obtained. Furthermore, from the definition of the function Ω, it follows that

$$W* = \Delta U - T\,\Delta S - \sum \Delta n_i\mu_i \tag{108}$$

which is the form in which Gibbs elected to introduce the work of nucleation [12]. It is important to note that $W*$ is clearly defined thermodynamically and that it is completely independent of the location of a dividing surface which separates the energy, entropy, and amounts of matter in the nucleus into surface and volume terms. It is also independent of any assumption as to the homogeneity of the nucleus.

To evaluate $W*$, Gibbs introduces the dividing surface and obtains the expression

$$W* = \sigma A_n - (p_n - p_o)V_n \tag{109}$$

which follows directly from equation (92). The following expression for size of the nucleus, if spherical in shape, is obtained from equation (89):

$$R_* = \frac{2\sigma_m}{p_n - p_o} \tag{110}$$

where the pressure difference $(p_n - p_o = \Delta p)$ represents the increase in pressure required to maintain bulk nucleus material in equilibrium with the exterior phase and σ_m is the surface tension at the surface of tension. By combining equations (109) and (110), the following three equivalent expressions for $W*$ are obtained:

$$W* = (16\pi/3)[\sigma_m{}^3/(\Delta p)^2] \tag{111}$$

$$W* = \tfrac{4}{3}\pi R_*{}^2\sigma_m \tag{112}$$

and

$$W^* = \tfrac{2}{3}\pi R_*{}^3 \, \Delta p \qquad\qquad (113)$$

According to equation (112), the minimum work of nucleus formation is equal to one-third of the work required to create the interface (surface of tension) and, by equation (113), it is numerically equal to one-half the work gained in forming the bulk phase from the exterior phase. W^* can be evaluated by equation (111) and R_* from either equation (112) or (113), if both σ_m and Δp are known. This is the case for large drops for which the radius of curvature is large compared to the thickness of the surface layer and for which the surface tension may be equated to that of a flat interface between the co-existing stable phases. Obviously, Δp will be very small and this assumption is strictly valid only when Δp approaches zero. When the droplet size decreases, the nucleus will be nonuniform throughout, and σ_m will be a function of its size. These conditions are to be expected in most nucleating systems, and equation (111) can no longer be used to evaluate W^*, but will serve merely as a definition of σ_m in terms of W^* and Δp; similarly, equation (113) will serve merely as a definition of R_*. This is the dilemma faced in nucleation theory. In the classical treatment of nucleation, regardless of the size of the nucleus, W^* is calculated by letting σ_m equal the surface tension of a plane liquid surface and by letting Δp equal the pressure between the two bulk phases when in equilibrium across a semipermeable barrier. Cahn and Hilliard [4] showed how the work W^* may be evaluated from a model which does not introduce surface tension, but which is based on an expression for the free energy of a nonuniform system. At very low supersaturations, the properties of the nucleus are shown to approach those predicted by the classical theory, which assumes the nucleus to be homogeneous and σ to be independent of curvature. This approach has the advantage over that of the classical theory in that it correctly predicts a continuous decrease in W^* with increasing supersaturation, with W^* approaching zero at the spinodal point.

It should be noted that Gibbs did not suggest the classical treatment of nucleation discussed above. He clearly realized that W^* must become zero at the spinodal point and that, according to equation (112), this can be accomplished only by a discontinuous change whereby W^*, σ_m, and R_* simultaneously reach zero values.

On the assumption of incompressibility of the liquid phase,

equation (106) when applied to single-component systems may be transformed to read

$$\mu_o(T, p_o) = \mu_n(T, p_o) + v_n(p_n - p_o) \tag{114}$$

where v_n is the molar volume of the liquid phase. Substitution of this equation into equation (111) gives the following alternative expression for W^*:

$$W^* = \left(\frac{16\pi}{3}\right) \frac{\sigma^3 v_n^2}{[\mu_o(p_o) - \mu_n(p_o)]^2} \tag{115}$$

The degree of supersaturation ($p_o/p_\infty = \alpha$) may be introduced into equation (115) by utilizing equation (97); thus,

$$W^* \cong \left(\frac{16\pi}{3}\right) \frac{\sigma^3 v_n^2}{(RT \ln \alpha)^2} \tag{116}$$

We note that equation (116) is based on the assumption that the difference $[\mu_n(p_o) - \mu_n(p_\infty)]$ is negligibly small. [See also the discussion preceding equation (99).]

Supersaturation may also be achieved by cooling the saturated vapor at pressure p_o and temperature T_o to temperature T, while maintaining constant pressure. To introduce the degree of undercooling $(T_o - T)$ in the expression for W^*, we consider the following reversible changes of state:

$$A \text{ (vapor, } p_o, T_o) = A \text{ (liquid, } p_o, T_o)$$

$$A \text{ (vapor, } p_o, T) = A \text{ (liquid, } p_n, T)$$

for which

$$\mu_o(p_o, T) - \mu_o(p_o, T_o) = \mu_n(p_n, T) - \mu_n(p_o, T_o)$$

and, therefore,

$$v_n(p_n - p_o) = -\Delta H \ln T/T_o \cong \frac{\Delta H(T_o - T)}{T_o} \tag{117}$$

where ΔH is the heat of evaporation (assumed independent of temperature over the temperature range of interest). The radius of the equilibrium nucleus is thus given by

$$R_m = \frac{2\sigma_m v_n T_o}{\Delta H(T_o - T)} \tag{118}$$

Substitution of this equation into equation (112) gives the following desired expression for W^*:

$$W^* = \left(\frac{16\pi}{3}\right) \frac{v_n^2 \sigma_m^3(T) T_o^2}{(\varDelta H)^2 (T_o - T)^2} \quad \vee \tag{119}$$

An expression for W^* identical to equation (112) may also be obtained by minimizing the expression for the work of formation W of a nucleus of arbitrary size R from the supersaturated parent phase of pressure p_o, subject to the additional constraints of constant pressure p_o, pressure of bulk nucleus material p_n, temperature T, and total number of moles N_i of component i. For a single-component parent phase, R_* is, therefore, determined by the following condition:

$$\delta(W)_{T, p_o, p_n, N} = 0 \quad \vee \tag{120}$$

where

$$W = -\frac{V}{v_n} [\mu_o(p_o) - \mu_n(p_o)] + \sigma A \quad \vee$$

$$W = -V \varDelta G_v + \sigma A \tag{121}$$

In equation (121), $\varDelta G_v$ denotes the Gibbs free energy change accompanying the formation of a unit volume of bulk nucleus material at pressure p_o. This quantity is related to the degree of supersaturation by the following approximation:

$$\varDelta G_v \cong RT \ln \alpha$$

[See equation (116).] Differentiation of equation (121) with respect to particle radius and application of equation (120) yields equation (110), which describes the radius of the equilibrium nucleus. From equations (110) and (121), it follows that

$$\left(\frac{d^2 W}{dR^2}\right)_{R=R_*} = -8\pi\sigma_m < 0$$

Therefore, W^* is a maximum and represents a barrier to homogeneous nucleation.

REFERENCES

1. J. D. van der Waals, *Verhandel. Koninkl. Ned. Akad. Wetenschap. Afdel. Natuurk. Sect. I* (1893).
2. G. Bakker, in: *Handbuch der Experimentalphysik*, Vol. 6, W. Wien and F. Harms (eds.), Akademische Verlagsgesellschaft (Leipzig), 1928, Chapter 15.
3. J. W. Cahn and J. E. Hilliard, *J. Chem. Phys.* **28**: 258 (1958).
4. J. W. Cahn and J. E. Hilliard, *J. Chem. Phys.* **31**: 688 (1959).
5. E. W. Hart, *Phys. Rev.* **113**: 412 (1959); *J. Chem. Phys.* **39**: 3075 (1963)
6. J. W. Gibbs, *Collected Works*, Vol. I, Yale University Press (New Haven, Connecticut), 1948, pp. 219–331.
7. J. W. Gibbs, *ibid.*, p. 219.
8. J. W. Gibbs, *ibid.*, p. 234.
9. J. W. Gibbs, *ibid.*, p. 223.
10. J. W. Gibbs, *ibid.*, p. 266.
11. J. W. Gibbs, *ibid.*, p. 253.
12. J. W. Gibbs, *ibid.*, p. 254.
13. J. W. Gibbs, *ibid.*, p. 300.
14. E. A. Guggenheim, *Trans. Faraday Soc.* **36**: 397 (1940).
15. S. Ono and S. Kondo, in: *Encyclopedia of Physics*, Vol. X, S. Flügge (ed.), Springer-Verlag (Berlin–Göttingen–Heidelberg), 1960, p. 144.
16. I. Langmuir, *J. Am. Chem. Soc.* **39**: 1848 (1917). See also N. K. Adam, *The Physics and Chemistry of Surfaces*, Oxford University Press (London), 1941.
17. W. D. Harkins, *The Physical Chemistry of Surface Films*, Reinhold Publishing Corp. (New York), 1952.
18. C. Herring, in: *Structure and Properties of Solid Surfaces*, R. Gomer and C. S. Smith (eds.), University of Chicago Press (Chicago), 1953, Chapter I.
19. W. W. Mullins, in: *Metal Surfaces*, American Society for Metals, 1963, Chapter 2.
20. N. Cabrera, *Surface Science* **2**: 320 (1964).
21. J. W. Cahn and J. E. Hilliard, *Acta Met.* **7**: 221 (1959).
22. R. S. Hansen, *J. Phys. Chem.* **66**: 410 (1962).
23. C. Herring, *Phys. Rev.* **82**: 87 (1951).
24. S. Ono and S. Kondo, *op. cit.*, p. 152.
25. R. Shuttleworth, *Proc. Phys. Soc. (London)* **A63**: 444 (1950).

Modern Theory of Fluid Surface Tension

Edward W. Hart

General Electric Research Laboratory
Schenectady, New York

INTRODUCTION

This paper presents a discussion of the spherical interface in fluid systems. The paper by de Bruyn* presented an exposition of the theory of surface tension due to J. W. Gibbs. I should like to emphasize that it is only in the context of that theory that the surface tension of equilibrium systems is properly described. Peculiarly enough, that theory has practically disappeared from the published literature and only a pale imitation of it is to be found now in the texts and treatises.

Although this subject was clearly and correctly presented by de Bruyn, it will be discussed here further. In particular, the simple case of a spherical interface in a one-component, two-phase system will be discussed in great detail, after which some new results that are not yet completely established, but imply some very interesting possibilities in this field, will be presented.

THE SPHERICAL INTERFACE

The subject of the spherical interface was treated specially by Gibbs [1], and what follows is essentially his treatment of the problem with some amplification.

The free energy function Ω was introduced elsewhere in this volume by de Bruyn. For a single-component system, it is defined by the following relation:

$$\Omega \equiv E - TS - \mu N \tag{1}$$

---

* See P. L. de Bruyn, "Some Aspects of Classical Surface Thermodynamics," this volume, pp. 1–36.

where E is the internal energy, S is the entropy, N is the number of moles, and T and μ are the temperature and chemical potential of the system, respectively. For any change of the state of the system along a reversible path, the mechanical work W done by the system is given by the following relation:

$$W = -\Delta\Omega \qquad (2)$$

For a simple homogeneous system,

$$\Omega = -PV \qquad (3)$$

where V is the volume of the system. We shall employ this function to describe the energetics of our problem, since it is that function appropriate to a system the temperature, chemical potential, and physical dimensions of which are chosen as the variables to specify.

Consider next the phase diagram in the $P–v$ plane of a one-component system as shown in Fig. 1. The isotherm shown has been extended into the interior of the phase-boundary locus. The "states" on that extension are not stable and represent experimentally attainable states that have macroscopic lifetimes. Because of this, we can describe them as equilibrium thermodynamic states, even though they are metastable. For definiteness, consider the state shown in Fig. 1 as (v_1, P_1).

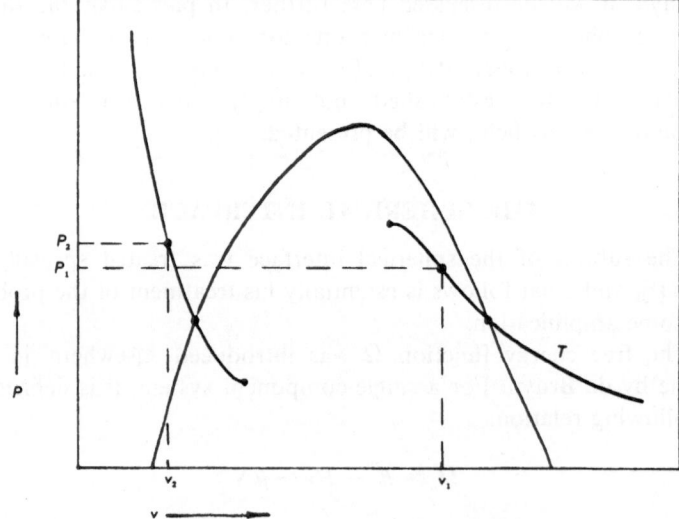

Fig. 1. A typical phase diagram for a simple one-component system. An isotherm is shown for temperature T. The variables are the pressure P and the molar volume v.

A chemical potential can be assigned to this state through integration of the Gibbs–Duhem equation from the phase boundary to the supersaturated state (v_1, P_1). Now it is not hard to see that there exists a condensed state (v_2, P_2) that has the same chemical potential. In general, P_2 is greater than P_1, and so the two phases cannot be placed in direct mechanical contact. It is conceivable, however, that there could exist a condensed phase similar to the bulk phase at (v_2, P_2) embedded in the phase (v_1, P_1) and in mechanical and chemical equilibrium. As a matter of fact, it is commonly believed that there is such an equilibrium (although unstable) state possible, e.g., a water droplet in equilibrium with supersaturated water vapor. We shall not examine the question of the possibility of the state any further, but shall simply point out that there is an assumption involved that has not yet had an unequivocal experimental demonstration. Furthermore, if we do not assume the existence of such a state, we cannot discuss the spherical interface at all! Gibbs was fully conscious of this state of affairs and explicitly discussed it [1].

We are to compute then the difference in Ω between two states— that of (v_1, P_1) alone, called state A, and that of a condensed droplet in vapor (v_1, P_1), called state B.

Regardless of the method of computation, the value of Ω in either state must be independent of the way in which the states are described. In other words, the value of Ω is a state property. The states are fully described by specifying the values of T, μ, and the volume V of the total system as well as the presence or absence of a droplet. Following Gibbs, we shall attempt to describe state B as though the droplet were the condensed phase (v_2, P_2), and we shall assign to it an arbitrary volume V_2, bounded by a spherical surface of area A. This is shown in Fig. 2. Write the value of Ω for state B as though it were the sum of Ω's for the two contiguous phases plus a remainder that will be written as proportional to A. Thus, we write

$$\Omega_A = -P_1 V \tag{4}$$

$$\Omega_B = -P_1(V - V_2) - P_2 V_2 + \sigma A \tag{5}$$

It must be emphasized at this point that the well-defined thermodynamic quantities in equations (4) and (5) are Ω_A, Ω_B, P_1, P_2, and V. The volume V_2 is arbitrary, and its choice determines the values of σ and A.

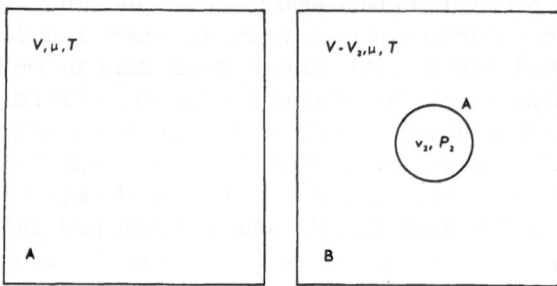

Fig. 2. Schematic representation of the states A and B.

Equations (4) and (5) may be combined in the following manner:

$$\Delta\Omega \equiv \Omega_B - \Omega_A$$
$$\dot{\Delta\Omega} = -(P_2 - P_1)V_2 + \sigma A$$
$$\Delta\Omega = -V_2\,\Delta P + \sigma A \qquad (6)$$

where $\Delta P \equiv P_2 - P_1$. This equation can be solved for σ;

$$\sigma = (\Delta\Omega + V_2\,\Delta P)/A \qquad (7)$$

or, if the radius of V_2 is R,

$$\sigma = (\Delta\Omega/4\pi)R^{-2} + (\Delta P/3)R \qquad (8)$$

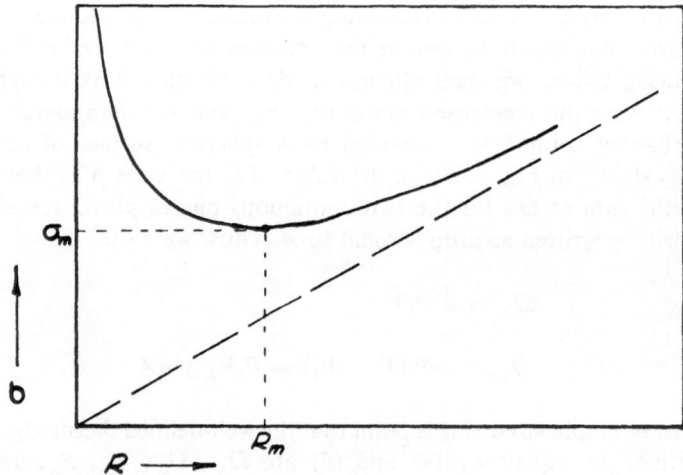

Fig. 3. A schematic plot of σ as a function of R, showing the position of the point for which $d\sigma/dR = 0$. The broken line through the origin corresponds to $\Delta\Omega = 0$.

Figure 3 shows a schematic plot of σ as a function of R. The meaning of σ is no more than is contained in equation (8). Since $\Delta\Omega$ is independent of the choice of R, the variation of equation (6) with respect to R must be zero. This leads to

$$0 = -4\pi R^2\, \Delta P + 4\pi R^2 (d\sigma/dR) + 8\pi R\sigma$$

and, finally, to the following equation:

$$\Delta P = \frac{2\sigma}{R} + \frac{d\sigma}{dR} \qquad (9)$$

This mechanical-looking equation exhibits the remainder term σA in equation (5) as behaving like the energy due to a membrane possessing a tension σ and some bending moments represented by $d\sigma/dR$. Figure 3 shows that R can be chosen so that $d\sigma/dR = 0$. The resultant value of σ that is called here σ_m is the tension of what Gibbs called the *surface of tension*. For that particular choice of fictitious surface radius,

$$\Delta P = \frac{2\sigma_m}{R_m} \qquad (10)$$

which is the familiar relationship quoted in all the literature.

As we have seen from the foregoing derivation, σ is a fairly arbitrary quantity. The usual treatment assumes that the value of σ that appears in equation (6) is well-defined *a priori* and that it satisfies equation (10) as well. It is then assumed to be the same as the surface tension of the planar interface, which, of course, corresponds to a different value of μ, and equation (6) then purports to be a computation of $\Delta\Omega$. Such a procedure is highly questionable and of little practical value.

Within the past ten years, an increasing quantity of research has been performed on the general subject of the thermodynamics and statistical mechanics of inhomogeneous equilibrium systems. The principal published literature in this field is by Cahn and Hilliard [2], Hart [3,4], Stillinger and Buff [5], and Lebowitz and Percus [6]. A recent application and further investigation of the results presented by Hart [4] lead to the conclusion that for the problem discussed above $\Delta\Omega = 0$. Although further investigation is in progress to test this result, a discussion of some of the consequences that are implied by it is presented below.

DISCUSSION OF RECENT RESULTS

When $\Delta\Omega = 0$, equation (8) may be rewritten in the following simple form:

$$\sigma = (\Delta P/3)R \qquad (8')$$

The plot of this function is, of course, a straight line and is shown in Fig. 3 as the broken line. The first conclusion is that, whatever the choice of R, $d\sigma/dR = \Delta P/3$, and so there does not exist a choice of R that will prescribe a surface of tension. Therefore, we cannot reduce equation (9) to the form of equation (10). This does not mean that the Gibbs theory is incorrect, but simply that there is no solution for the equation $d\sigma/dR = 0$.

In the nucleation theory of the kinetics of condensation of supersaturated vapor, the rate of nucleation $\mathscr{R}$ is dependent on what is called the free energy of activation in the form

$$\mathscr{R} = \nu e^{-(Q/kT)} \qquad (11)$$

In the usual treatment, the principal part of Q is the quantity $\Delta\Omega$ that we have been discussing. If $\Delta\Omega = 0$, the rate $\mathscr{R}$ would no longer satisfy an Arrhenius relation of the form of equation (11). Nevertheless, experimental observations confirm the Arrhenius-type behavior. The conclusion that would be required by our present analysis is that Q refers to the activation free energy of a state that has not yet been identified and that occurs on the reaction path somewhere before the appearance of the state B that we have discussed here. That true activated state cannot then be characterized by the same value of μ as the states A and B.

REFERENCES

1. J. W. Gibbs, *The Scientific Papers*, Vol. I, Dover Publications, Inc., (New York), p. 252 *et seq.*
2. J. W. Cahn and J. E. Hilliard, *J. Chem. Phys.* **28**: 258 (1958).
3. E. W. Hart, *Phys. Rev.* **113**: 412 (1959); **114**: 27 (1959).
4. E. W. Hart, *J. Chem. Phys.* **39**: 3075 (1963).
5. F. H. Stillinger, Jr., and F. P. Buff, *J. Chem. Phys.* **37**: 1 (1962).
6. J. L. Lebowitz and J. K. Percus, *J. Math. Phys.* **4**: 116 (1963).

Nucleation in Homogeneous Vapors

K. C. Russell

Massachusetts Institute of Technology
Cambridge, Massachusetts

INTRODUCTION

This is essentially a review of a longer, more rigorous paper by J. Feder, K. C. Russell, J. Lothe, and G. M. Pound [6]; those desiring a more complete discussion of the topics in this paper are referred to this work of Feder *et al.*

Nucleation theory may be said to have originated with the formulation of the following expression for the vapor pressure over a spherically curved liquid surface:

$$\ln p_r/p_\infty = \frac{2\sigma V_B}{rkT} \tag{1}$$

p_r is the pressure over the curved surface; p_∞, the pressure over a planar surface; σ, the liquid–vapor surface tension; V_B, the molecular volume in the liquid; r, the radius of curvature; and kT, the product of Boltzmann's constant and the absolute temperature.

Any unary droplet is unstable with respect to a planar surface, but it also may be unstable with respect to growth, unstable with respect to evaporation, in unstable equilibrium, or metastable in a system supersaturated with respect to the bulk material. The supersaturation S is the ratio of the actual and saturation vapor pressures in a system. The cluster in unstable equilibrium with a vapor of given supersaturation is known as the critical nucleus and plays a vital part in the theory of nucleation. Gibbs [8] realized this and suggested $e^{-w^*/kT}$ (where w^* is the reversible work to create a critical nucleus from the supersaturated vapor) as a criterion for the stability of a supersaturated vapor.*

---

* Note that an asterisk denotes quantities taken at the critical size.

He gave $w^* = \frac{1}{3}\sigma A^*$, where A^* is the surface area of the critical nucleus. The radius corresponding to A^* may easily be obtained by inverting equation (1):

$$r^* = \frac{2\sigma V_B}{kT \ln S} \tag{2}$$

where r^* is infinite for a saturated vapor ($S = 1$) and negative (meaningless) for undersaturated vapors ($S < 1$). The term $e^{-w^*/kT}$ changes from an extremely small number ($\sim e^{-20,000}$) to a more reasonable quantity ($\sim e^{-50}$) as r^* goes from 1000 Å to, e.g., 5 Å.*

Although $e^{-w^*/kT}$ is a good estimate of the barrier to nucleation, it lacks the pre-exponential and frequency factors needed in a nucleation rate equation. Methods of obtaining these factors will be discussed, as will the controversy over the thermodynamics of clusters in a supersaturated vapor.

A simple derivation of the time lag for steady state obtained by Feder et al. [6] will be discussed with the underlying irreversible thermodynamic principles. Finally, theory and experiment are compared and suggestions made for future experimental work.

KINETIC EQUATION

Much of the work on nucleation since Gibbs has involved the determination of the pre-exponential factors accompanying $e^{-w^*/kT}$. Two methods are used for this: one, called the kinetic treatment, uses the vapor pressure over a curved surface; and the other, called the pseudo-equilibrium treatment, involves the metastable equilibrium concentration of clusters. It is a common belief that the two treatments differ in principle. Actually, the two treatments are fundamentally the same.

A brief outline of the two methods of deriving the steady-state flux of clusters will show the equivalence. Nucleation theory uses cluster size as a configurational coordinate. A flux J in this configurational space is the number of clusters/cm³-sec going from size n to $n + 1$. The net flux between n and $n + 1$ is given by

$$J_n = \beta A(n)c(n) - \alpha(n + 1)A(n + 1)c(n + 1) \tag{3}$$

where $\beta = p/\sqrt{2\pi mkT}$ is the impingement frequency of monomer on a planar surface; $\alpha(n + 1)$ is the evaporation rate per unit area of a

---

* One notes the difficulty in ascribing the macroscopic σ to clusters of 5 Å.

cluster containing $n + 1$ molecules; $c(n)$ is the number of clusters of size n per cubic centimeter; and $A(n)$ is the surface area of a cluster of n molecules. The metastable equilibrium treatment invokes the principle of detailed balance, which states that at equilibrium every microscopic process occurs at the same rate as its inverse. For example, in a vapor at equilibrium, the number of clusters growing from n to $n + 1$ is exactly balanced by those decomposing from $n + 1$ to n. If cluster–cluster collisions are considered, at equilibrium each forward reaction must be exactly balanced by its inverse. For example, if there is a finite probability of 25-mer (clusters containing 25 molecules) colliding to form 50-mer, then 50-mer must decompose to 25-mer at a rate to balance out the forward reaction.

Detailed balance is applied to nucleation in supersaturated systems by assuming some artifice—a Maxwell demon—to keep supercritical clusters from growing and causing a net flux. If this is permitted, one may take the fluxes as zero at equilibrium and write the following relationship:

$$J_n = 0 = \beta A(n)c_0(n) - \alpha(n + 1)A(n + 1)c_0(n + 1)$$

where the subscript 0 indicates a quantity considered at equilibrium. We may now evaluate the following expression:

$$\alpha(n + 1) = \frac{\beta A(n)c_0(n)}{A(n + 1)c_0(n + 1)} \tag{4}$$

If $\alpha(n + 1)$ is the same whether or not equilibrium exists, one may write

$$J_n = \beta A(n)c(n) - \frac{\beta A(n)c_0(n)}{A(n + 1)c_0(n + 1)} A(n + 1)c(n + 1) \tag{5}$$

Approximation of the difference $c(n)/c_0(n) - c(n + 1)/c_0(n + 1)$ as a differential yields

$$J = -\beta A(n)c_0(n) \frac{\partial [c(n)/c_0(n)]}{\partial n} \tag{6}$$

Although $c(n)$ and $c_0(n)$ vary greatly over one size interval, their ratio changes hardly at all. The first-order expansion is proper for c/c_0, whereas one would need a lengthy series to express either $c(n) - c(n + 1)$ or $c_0(n) - c_0(n + 1)$ in terms of derivatives.

The vital assumption here is that α is the same in a nonequilibrium situation as at equilibrium. This assumption has been attacked repeatedly, but it is absolutely necessary for the development of a quantitative theory of nucleation. Even in cases where very high fluxes ($J > 10^8$ clusters/cm³-sec) exist, this assumption seems valid since the difference between the forward and back fluxes is still very small. We will see later that, proportionately, the greatest net flux exists near the critical region, and even here the forward flux is only a few percent greater than the reverse. In the lower size classes, the net flux is a very small fraction of the total—of the order of $10^{-8}\%$. Therefore, even at high fluxes, the clusters in the various size classes have nearly as much time to obtain the equilibrium configuration as if true equilibrium existed.

The term $c_0(n)$ is generally expressed as

$$c_0(n) = c(1) \exp\left(-\Delta G_n^0/kT\right) \tag{7}$$

with

$$\Delta G_n^0 = -nkT \ln p/p_\infty + \sigma A(1)n^{2/3} \tag{8}$$

where $A(1)$ is defined by $A(n) = A(1)n^{2/3}$; the term p/p_∞ is the supersaturation ratio of the system; σ is the surface energy of the liquid–vapor interface; and ΔG_n^0 shows a maximum at n^*;

$$\left(\frac{\partial \Delta G_n^0}{\partial n}\right)_{n^*} = -kT \ln p/p_\infty + \tfrac{2}{3}\sigma A(1)n^{-1/3} = 0$$

and

$$n^* = \left[\frac{2\sigma A(1)}{3kT \ln p/p_\infty}\right]^3 \tag{9}$$

Equation (9) is equivalent to the expression derived earlier for r^*:

$$r^* = \frac{2\sigma V_B}{kT \ln p/p_\infty} \tag{2'}$$

where r^* and n^* are related by the equation $V_B n^* = \tfrac{4}{3}\pi r^{*3}$ and V_B is the liquid molecular volume.

Since $c_0(n)$ shows a minimum at n^*, one suspects that this region will throttle the nucleation process and that what occurs in the lower size classes is of little importance.

This is indeed the fact, as was shown in an ingenious evaluation of the steady-state flux performed by Farkas [5]. Steady state is obtained when the flux is independent of the size class, and this constancy may be used to express the flux in terms of supersaturation ratio, temperature, and other measurable quantities.

Equation (6) may be rewritten as

$$\frac{J \, dn}{\beta A(n)c_0(n)} = -d(c/c_0) \tag{10}$$

which may be integrated subject to the following boundary conditions:

$$c/c_0 = 1 \quad \text{at} \quad n = 1$$

and

$$c/c_0 = 0 \quad \text{as} \quad n \to \infty$$

The first boundary condition means that the monomer pressure is fixed at the equilibrium value, either by walls permeable to monomer or by comminution of the supercritical clusters to monomer. The other boundary condition indicates that clusters much larger than n^* are removed from the system. How much larger is immaterial, since clusters with $n \gtrsim 2n^*$ have a negligible probability of decomposing. We now have the following relationship:

$$\int_1^\infty \frac{J \, dn}{\beta A(n)c_0(n)} = -\int_1^0 d(c/c_0) \tag{11}$$

Only $A(n)$ and $c_0(n)$ vary with n in the integrand on the left. Furthermore, $c_0(n)$ depends exponentially on n, and, thus, the $n^{2/3}$ factor in $A(n)$ is relatively insignificant. Since $c_0(n)$ has a minimum at n^*, $1/c_0(n)$ will be a maximum in that region. Most of the value of the integral comes from the immediate neighborhood of n^*, and one may utilize a power series in the evaluation. It is convenient to use an exponential power series in this case, expanding ΔG_n^0 rather than $c_0(n)$. Figure 1 is a plot of ΔG_n^0 vs. n.

One has

$$\Delta G_n^0 = -nkT \ln p/p_\infty + \sigma A(1)n^{2/3} \tag{8}*$$

$$\left(\frac{\partial^2 \Delta G_n^0}{\partial n^2}\right)_{n*} = -\tfrac{2}{9}\sigma A(1)n^{*-4/3} \tag{12}$$

and

$$\Delta G_n^0 \cong \Delta G^* - (\tfrac{2}{3})\sigma A(1)n^{*-4/3}\frac{(n - n^*)^2}{2} \tag{13}$$

* σ is assumed equal to that for a plane surface.

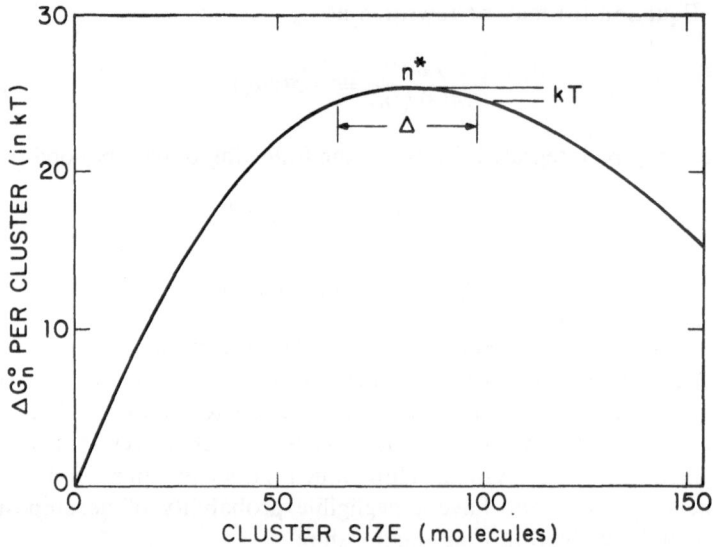

Fig. 1. Standard free energy change per cluster ΔG_n^0 as a function of cluster size for water vapor at 273°K and a supersaturation of 4. (From $n = 20$ to $n = 1$, a smooth curve is drawn to make $\Delta G_1^0 = 0$.) Taken from Feder *et al.* [6].

since

$$\left(\frac{\partial \Delta G_n^0}{\partial n}\right)_{n*} = 0$$

One may approximate $A(n)$ as $A(n^*)$ and remove from the integral that factor as well as the constants J, β, and $c_0(n^*)$, which leaves

$$\frac{J}{\beta A(n^*)c_0(n^*)} \int_1^\infty dn \exp\left[\frac{1}{kT}\left(\frac{\partial^2 \Delta G_n^0}{\partial n^2}\right)_{n*} \frac{(n - n^*)^2}{2}\right] \quad (14)$$

Since the exponential has negligible value far from n^*, we may expand the region of integration to $\pm\infty$ and replace dn by $d(n - n^*)$. The integral is then of standard form and yields

$$J = \left[\frac{\sigma A(1)}{9\pi k T n^{*4/3}}\right]^{\frac{1}{2}} \beta A(n^*)c_0(n^*) \quad (15)$$

where the term $[\sigma A(1)/9\pi k T n^{*4/3}]^{\frac{1}{2}}$ sometimes is denoted by Z and referred to as the Zeldovich factor. The presence of the $c_0(n^*)$ term in equation (15) sometimes leads to the erroneous belief that this is an

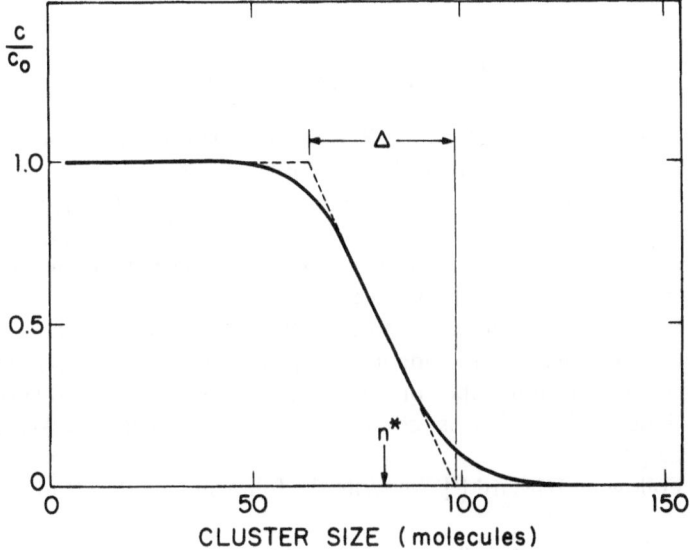

Fig. 2. Steady-state—divided by equilibrium—concentrations of clusters in size as a function of cluster size for water vapor at 273°K and a supersaturation of 4. The width Δ of the critical region is indicated. Taken from Feder et al. [6].

equilibrium theory. Such is not the case, since the theory is for steady state, and the Zeldovich factor may be considered to account for the depletion of critical clusters below the equilibrium value and for the decomposition of supercritical clusters.

Figure 2 is a plot of c/c_0 vs. n at steady state, where it can be seen that $c \cong c_0$ (i.e., $c/c_0 = 1$) almost up to the value of n^*. At n^*, $c = c_0/2$, as must be the case for a diffusional process of this sort. The gross rate at which critical clusters are promoted to $n^* + 1$ is $J_G{}^* = D^*c^*$, and the net flux is

$$J^* = ZD^*c_0(n^*)$$

The transmission coefficient K for passage across the critical region is

$$K = \frac{J^*}{J_G} = Z\left(\frac{c_0}{c}\right)$$

which is of the order of 10^{-1}. In the lower size classes, $J = J^*$, and

$$J_G = D(n)c(n) \cong D(n)c_0(n)$$

and

$$K = \frac{ZD^*c_0(n^*)}{D(n)c_0(n)} \cong \frac{c_0(n^*)}{c_0(n)}$$

If n is appreciably less than n^*, K is a very small number indeed, and the back flux is nearly identical to the forward flux.

We note that J depends mainly on ΔG^* [through $c_0(n^*)$] and weakly on the shape of ΔG_n^0 near n^*. The behavior of $c_0(n)$ in the lower size classes is, therefore, of no importance, and the critical region indeed throttles the nucleation of droplets.

It is a fairly simple matter to show that the equilibrium theory just given is equivalent to the kinetic theory based on equation (1), which gives the equilibrium vapor pressure over a spherically curved surface. The kinetic derivation is commonly associated with Becker and Döring [2].

Following Barnard [1], we write the following for the steady state:

$$\begin{aligned}
J &= \beta A(1)c(1) - \alpha_2 A(2)c(2) \\
J &= \beta A(2)c(2) - \alpha_3 A(3)c(3) \\
&\vdots \qquad\qquad \vdots \\
J &= \beta A(N-1)c(N-1) \qquad \text{for} \quad N \gg n^*
\end{aligned} \qquad (16)$$

where α_n is the vaporization frequency per unit area of an n-mer. Note that $\beta A(n)c(n)$ is the rate n-mer is promoted to $(n+1)$-mer and $\alpha_{n+1}A(n+1)c(n+1)$ is the rate of the reverse process. We have set $c(N) = 0$ equivalent to the upper boundary condition for the integral in equation (11). One may eliminate the c's from these equations by multiplying the second by α_2/β, the third by $\alpha_2\alpha_3/\beta \cdot \beta$, etc. The equations are added to give the following:

$$\frac{\beta A(1)c(1)}{J} = 1 + \frac{\alpha_2}{\beta} + \frac{\alpha_2\alpha_3}{\beta^2} + \cdots + \frac{\alpha_2\alpha_3 \cdots \alpha_{N-1}}{\beta^{N-2}} \qquad (17)$$

Whereas α was eliminated in favor of c_0 before, we have now eliminated c in favor of α. To proceed further, it is necessary to assume that the vapor pressure (or rather vaporization frequency of the clusters) is the same at steady state as at equilibrium. This is equivalent to the assumption of detailed balance in obtaining α in the previous derivation; thus, the kinetic derivation is also a near-equilibrium treatment and subject to the same objection as the pseudo-equilibrium theory.

Leaving Barnard, we may write

$$\Delta\mu_n = \frac{\partial \Delta G_n{}^0}{\partial n} = kT \ln p_n/p \tag{18}$$

where $\Delta\mu_n$ is the chemical potential of the molecules within an n-mer, referred to the supersaturated vapor; p_n is the vapor pressure of the droplet; and p is the pressure of the supersaturated vapor. One notes that $p_{n^*} = p$ and $\Delta\mu_{n^*} = 0$. Thus,

$$\frac{p_n}{p_{n^*}} = \exp\left(\frac{\Delta\mu_n}{kT}\right)$$

Use of the supersaturation ratio

$$\frac{p_{n^*}}{p_\infty} = p/p_\infty = S$$

yields

$$\frac{\alpha_n}{\beta} = \exp(\Delta\mu_n/kT)$$

and the nth term in equation (17) may be written as follows:

$$\frac{\alpha_2\alpha_3\alpha_4 \cdots \alpha_n}{\beta^{n-1}} = \exp\left(\frac{1}{kT}\sum_2^n \Delta\mu_n\right) \tag{19}$$

Replacement of the sum by an integral yields

$$\exp\left(\frac{1}{kT}\sum \Delta\mu_n\right) \cong \exp\left(\frac{1}{kT}\int_2^n \Delta\mu_n \, dn\right) \tag{20}$$

The integral of a chemical potential with respect to amount of material is a free energy; thus, if we take

$$\Delta\mu_n = -kT \ln S + \tfrac{2}{3}\sigma A(1)n^{-1/3} \tag{21}*$$

the integral becomes

$$\exp\left(\frac{1}{kT}\int_2^n \Delta\mu_n \, dn\right) = \exp\left(\frac{+\Delta G_n{}^0}{kT}\right) \tag{22}$$

* Any form of $\Delta G_n{}^0$ will be restored by the integration, as we are only integrating a differential. Specifically, the statistical-mechanical terms discussed later are automatically included in $\Delta G_n{}^0$ if one uses the proper expression for $\Delta\mu_n$.

We have ignored the difficulty at the lower limit of integration, since this is indicative of a problem to be discussed later in thermodynamics of clusters. We now have

$$\frac{\beta A(1)c(1)}{J} = \sum_0^{N-2} \exp\left(\frac{+\Delta G_n^{\,0}}{kT}\right) \tag{23}$$

This is seen to be very similar to equation (11), and indeed the sum may be converted to an integral and evaluated by an exponential power series, which yields

$$J = \left[\frac{\sigma A(1)}{9\pi k T n^{*4/3}}\right]^{\frac{1}{2}} \beta A(1)c_0(n^*) \tag{24}$$

This differs from equation (15) in the appearance of $A(1)$ in place of $A(n^*)$ in the pre-exponential. This is a minor difference arising from difficulties in changing from differences to differentials.

Thus, the two treatments are shown to be equivalent. The first derivation is to be preferred, however, as it deals primarily with the region near n^* and avoids the discreteness difficulty which arises near $n = 1$.

THERMODYNAMICS

The expression for the equilibrium distribution in size was taken earlier to be·

$$c_0(n) = c(1) \exp\left[n \ln S - \frac{\sigma A(n)}{kT}\right] \tag{25}$$

This is a violation of the law of mass action [6] and cannot possibly be correct. The term $n \ln S$ may be taken out of the exponent to give

$$c_0(n) = (c_1) \left(\frac{c_1}{c_{1\infty}}\right) \exp\left[-\frac{\sigma A(n)}{kT}\right] \tag{26}$$

where $c_{1\infty}$ is the concentration of monomer in the vapor just saturated with respect to bulk liquid. The term σ is taken independent of monomer pressure. Then, at constant temperature,

$$c_0(n)\alpha(c_1)^{n+1} \tag{27}$$

This is an impossibility, as one must have $c_0(n)\alpha(c_1)^n$ for a reaction of the following type:

$$nA \rightleftarrows A_n \tag{28}$$

Further, a cluster of n molecules has been assumed to differ from an equivalent amount of bulk material only by the presence of the surface with planar surface energy. This is only a first-order approximation, since the clusters behave as macromolecules and are free to translate and rotate throughout the volume of the vapor. This problem has been attacked several times—first by Volmer [17], later by Frenkel [7], Rodebush [14,15], Kuhrt [11], and Lothe and Pound [12]. The last paper established the importance of the external coordinates of the clusters, although the other papers were reasonably correct in estimating the size and form of the effect.

Perhaps the most lucid derivation of the equilibrium distribution of clusters is obtained by minimizing the Helmholtz free energy of a system containing all possible clusters subject to constraints of constant temperature, volume, and amount of material. The following derivation is from the work of Feder *et al.* [6] and generally follows that of Kuhrt [11].

Consider a box of unit volume and constant mass of monomer in a heat bath at temperature T. Let $c(n)$ denote the number of clusters of size n. Considering the system of clusters as a mixture of ideal gases, we may write the Helmholtz free energy of the system of clusters as follows:

$$F = \sum_{n=1}^{\hat{n}} F(n) = \sum_{n=1}^{\hat{n}} \left[c(n)f(n) - c(n)kT \ln \frac{e}{c(n)} \right] \qquad (29)$$

where e is the natural logarithm base; $\hat{n}$ is the largest cluster allowed to exist; $F(n)$ is the free energy of each subsystem of $c(n)$ clusters; and $f(n)$ is the free energy of a single cluster (the last two terms both refer to unit volume). One may minimize F with respect to the number of clusters in the various size classes subject to the constraint of constant mass and obtain the following relationship:

$$c_0(n) = \exp\{-[f(n) - n\mu_v]\}/kT \qquad (30)$$

where μ_v is the chemical potential of the vapor. The problem is reduced to finding $f(n)$, the free energy of an n-mer in a box of unit volume. Feder *et al.* [6] estimate

$$f(n) = nf_0 - \frac{3kf_T}{c_l} + \sigma A(n) - kT \ln \frac{aV_B n^{5/2}}{\lambda^3} - kT \ln \frac{n^{3/2}}{\lambda^3} \qquad (31)$$

where a is a geometric factor equal to 4.76 for a spherical drop; nf_0 is the free energy of n atoms of bulk liquid; and where

$$\lambda = \frac{h}{\sqrt{2\pi mkT}}$$

The arguments of the logarithms are the rotational and translational partition functions of the cluster in a box of unit volume, and the last two terms account for the contributions of this motion to the free energy. The free energy of a cluster is the free energy of the equivalent bulk corrected for surface $[\sigma A(n)]$ and for motion of the droplet as a whole (the logarithmic terms). The term $3kf_T/c_l$, where f_T is the thermal free energy of a liquid molecule (excluding the binding energy) and c_l is the molecular specific heat of the liquid, serves to conserve degrees of freedom. The six external degrees of freedom (three rotational and three translational) replace six vibrational modes in the liquid, and this term is intended to account for the effect. As written, this replacement term is a minor quantity of some $5kT$.

We may write

$$\mu_v = \mu_\infty + kT \ln p/p_\infty$$

and take $\mu_\infty \sim f_0$. Minimizing F subject to constant T, V, and mass, we find

$$c_0(n) = \frac{aV_B n^4}{\lambda^6} \exp\left(\frac{3f_T}{c_l T}\right) \exp\left(n \ln S - \mu n^{2/3}\right) \qquad (32)$$

where

$$\mu = \frac{\sigma A(1)}{kT}$$

Here, μ is a reduced surface tension and not a chemical potential. First, one notes that the spurious $c(1)$ term has disappeared from the pre-exponential. (The term $aV_B n^4/\lambda^6$ has units of cm^{-3}, however.) The exponent is the same as before, but the pre-exponential is some $10^{18}c(1)$ for typical conditions. Since the steady-state flux of clusters is proportional to $c_0(n^*)$, use of the equilibrium distribution just given yields fluxes 10^{18} times higher than before at the same temperature and pressure. Figure 3 shows steady-state flux *versus* supersaturation ratio for the classical and Lothe–Pound theories.

Such a change has naturally aroused controversy, and most of the argument has centered on the replacement term [6]. Some, following

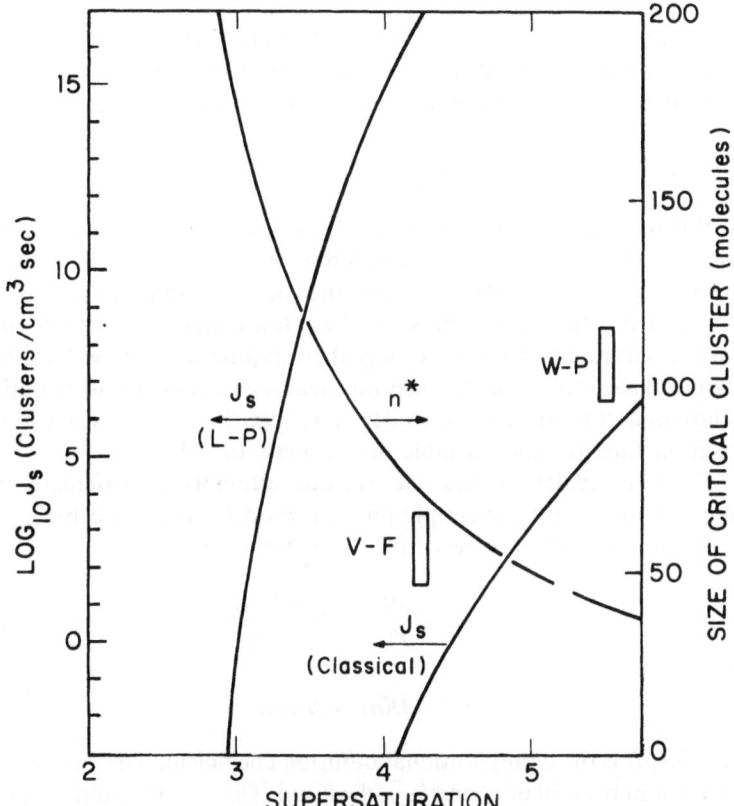

Fig. 3. Steady-state nucleation rate *vs.* supersaturation for water vapor at 273°K.
L–P refers to the theory of Lothe and Pound [12]. W–P and V–F refer to the data
of Wilson [19] and Powell [13] and Volmer and Flood [18], respectively. Classical
refers to the theory of Frenkel [7]. Taken from Feder *et al.* [6].

Kuhrt [11], wish to include the liquid binding energy in f_T, thus making
the increase in $c_0(n)$ some 10^8 rather than 10^{18}. Dunning [4] has devised
a statistical-mechanical derivation wherein $c_0(n)$ is only a factor of 10^4
greater than the classical value. He acknowledges the existence of
translation and rotation of the droplets in the vapor, but he ascribes
nearly the same degree of rotational and translational freedom to
equivalent masses of material in the bulk liquid. The translation and
rotation in the vapor phase are, thus, nearly cancelled by equivalent
movement in the liquid phase. This procedure seems physically
unrealistic, but controversy over the correction factor is still active

at the time of this writing. Feder *et al.* [6] believe the factor of 10^{18} to be essentially correct, and they advance supporting arguments in terms of thermodynamic cycles and broken-bond models.

DELAY TIME

When a vapor is suddenly made supersaturated, one presumes that virtually all the material is present as monomer. The number of *n*-mer in a vapor varies as $[c(1)]^n$; therefore, the concentration of, e.g., 50-mer is a very strong function of pressure. Very few clusters will be inherited from the saturated vapor. One may then inquire as to how long this system will need to establish something near the steady-state distribution of clusters and therewith the steady-state flux. In essence, what is the relaxation time for the unstable, initial mass distribution?

Feder *et al.* [6] discuss the various attempts to calculate this quantity and present the simple physical model reproduced here.

Equation (6) may be rewritten in the following form:

$$J = -D \frac{\partial c}{\partial n} + Dc \frac{\partial \ln c_0}{\partial n} \tag{33}$$

with

$$D = D(n) = \beta A(n)$$

where $\beta A(n)$ is the configurational diffusion coefficient. The dependence of c on n and t and of c_0 on n is understood. This is a diffusion equation wherein the first term represents random walk and the second represents drift in a potential gradient of $\partial \ln c_0/\partial n$ or $-\partial(\Delta G_n^0/kT)/\partial n$. Figure 1 shows this potential field.

Knowing that the nucleation process is throttled by the region near n^*, one may expect a sigmoidal dependence on time of the critical cluster flux. Such a behavior may be expressed analytically by equations of the form

$$J^*(t) = J_{ss} \exp(-\tau/t) \tag{34a}$$

or

$$J^*(t) = J_{ss}[1 - \exp(-t/\tau)] \tag{34b}$$

where $J^*(t)$ is the time-dependent flux of critical clusters; J_{ss} is the steady-state flux, the same for all size classes; and τ, in both cases, is the time needed for the flux of critical clusters to reach an appreciable fraction ($\sim 1/e$) of the steady-state value. The system is presumed to be essentially pure monomer at $t = 0$, but small concentrations of clusters do not alter the treatment.

Most efforts to obtain τ have been mathematical attempts to solve the time-dependent nucleation equation. This is obtained by applying the continuity relation to equation (33) [20]:

$$\frac{\partial c}{\partial t} + \frac{\partial J}{\partial n} = 0$$

or

$$\frac{\partial c}{\partial t} = \frac{\partial}{\partial n} \left(D \frac{\partial c}{\partial n} - Dc \frac{\partial \ln c_0}{\partial n} \right) \tag{35}$$

Feder et al. [6] chose physics instead of mathematics to obtain their delay time τ. Instead of attempting to deduce the time dependence of equation (35), they considered the behavior of individual clusters, reasoning that the average travel time taken by a critical cluster to pass from $n = 1$ to $n = n^*$ is a good measure of the delay time. This is, in fact, so, since the clusters interact only with monomer, not with each other. Thus, a cluster will move about n-space in the same manner whether the system is in a transient, steady state, or in an equilibrium condition.

The growth of a cluster from monomer is essentially a fluctuation and, thus, a very unlikely process. Since many orders of magnitude more dimer, trimer, etc., are formed than n^*-mer, it is very difficult to deduce the average path or pattern of cluster growth, since one does not know which cluster to follow. To find this average behavior, one needs the principle of time reversal, i.e., fluctuations decay in the same way they form. This is the principle used by Onsager to derive his celebrated reciprocal relations; and, while these relations are of little use in nucleation, time reversal is of immense value.

Since the formation of critical cluster is a fluctuational process, we may use our knowledge of the decomposition to infer the mode of growth. Consider a critical cluster which (by definition) has the same probability of growing or decomposing. The drift-velocity term in equation (35) is zero since $\partial \ln c_0/\partial n = 0$; therefore, the cluster performs a random walk near n^* (see Fig. 2). Should the cluster become as much as 10 or 20 units super- or subcritical, it would be appreciably unstable and would tend to grow or evaporate, respectively. In either case, the cluster has moved off the flat part of the potential and movement is controlled by drift in the potential gradient and the random-walk motion is unimportant (see Fig. 1).

The width of the random-walk region may be estimated several ways [6], but perhaps the simplest is to take $\Delta = \frac{1}{Z}$, where Z is the Zeldovich factor in equation (15). Figure 2 shows c/c_0 vs. n at the steady state. Since

$$J^* = D^* c_0(n^*) \left(\frac{\partial c/c_0}{\partial n} \right)_{n^*}$$

one sees that

$$Z = \left(\frac{\partial c/c_0}{\partial n} \right)_{n^*}$$

See equation (15). Also, c/c_0 is almost linear over the region of width Z, which indicates that movement of clusters in this region is essentially by random walk. In regions where c/c_0 equals one or zero, growth is controlled by the drift velocity. Only in very small intervals near $n^* - \frac{\Delta}{2}$ and $n^* + \frac{\Delta}{2}$ are the types of growth mixed, and $\frac{1}{Z}$ is an excellent measure of the essentially flat region where random walk dominates.

One should not consider a cluster to be nucleated until there is only a small chance of re-evaporation. For this reason, Feder et al. [6] consider a nucleus of size $n^* + \frac{\Delta}{2}$ to be stable and capable of sustained growth. The delay time, then, is the time required for a cluster to reach $n^* + \frac{\Delta}{2}$ from $n = 1$. We calculate the time for a cluster to decompose from $n^* + \frac{\Delta}{2}$ to $n = 1$ by separating the path into two parts—that between $n^* + \frac{\Delta}{2}$ and $n^* - \frac{\Delta}{2}$ and that between $n^* - \frac{\Delta}{2}$ and $n = 1$. From elementary diffusion theory, the time required for a cluster to random walk a distance Δ is

$$t_1 = \frac{\Delta^2}{2D^*} \tag{36a}$$

or

$$t_1 = \frac{1}{2D^* Z^2} \tag{36b}$$

The time required for a cluster to drift from $n^* - \frac{\Delta}{2}$ to 1 is

$$t_2 = \int_{n^* - \frac{\Delta}{2}}^{1} \frac{dn}{\dot{n}} \tag{37}$$

where the drift velocity is

$$\dot{n} = D \frac{\partial \ln c_0(n)}{\partial n}$$

Use of equation (8) for $c_0(n)$ yields

$$\dot{n} = D \left(\ln S - \frac{2}{3}\mu n^{-1/3} + \frac{4}{n} \right) \tag{38}$$

The $\frac{4}{n}$ term is from the translational and rotational contributions to $c_0(n)$. Integration and simplification yield

$$t_2 = \frac{n^{*4/3}}{D^*\mu} \ln \left(\frac{4\mu}{9\pi e^2} n^{*2/3}\right) \tag{39}$$

The average time required for a cluster to evaporate from $n^* + \frac{4}{2}$ to monomer is $\tau = t_1 + t_2$. By the principle of time reversal, this is, on the average, the *same time* taken to reach $n^* + \frac{4}{2}$ from monomer; thus, the delay time for steady state is $\tau = t_1 + t_2$. Under most conditions, with n^* of the order of 100, $t_1 > 2t_2$; therefore, we take

$$\tau \cong t_1 = \frac{1}{2D^*Z^2}$$

which may also be expressed as follows:

$$\tau = \frac{9\pi n^{*4/3}}{2D^*\mu} \tag{40}$$

This is consistent with the better mathematical derivations of τ. One recalls that here $D^* = \beta A(n^*)$ and $\mu = \sigma A(1)/kT$.

Time reversal is essential in this case, since the average time for a cluster to diffuse *up* a potential gradient is negative and meaningless. We are interested only in those clusters which reach the top, not the average dimer, trimer, *etc.*, which almost always evaporate. We could not use the drift-velocity calculation at the top of the barrier, because here the velocity is zero and the time of passage is infinite. Had we assumed that the clusters moved all the way from monomer by random walk, the value of τ would have been some two orders of magnitude too great. As it is, for water vapor, typically, $n^* \cong 100$, $D^* \cong 10^8$, $\mu \cong 10$, and $\tau \cong 10^{-6}$ sec. This method of estimating time lags could well be applied to other nucleation problems without excessive difficulty.

COMPARISON WITH EXPERIMENT

Two factors, which were dealt with by Feder *et al.* [6], could complicate comparison of theory and experiment—release of latent heat in the growing nuclei and consumption of monomer by growing clusters.

When a molecule adds on to a cluster, the latent heat is released. This raises the temperature and evaporation rate of the cluster, thereby

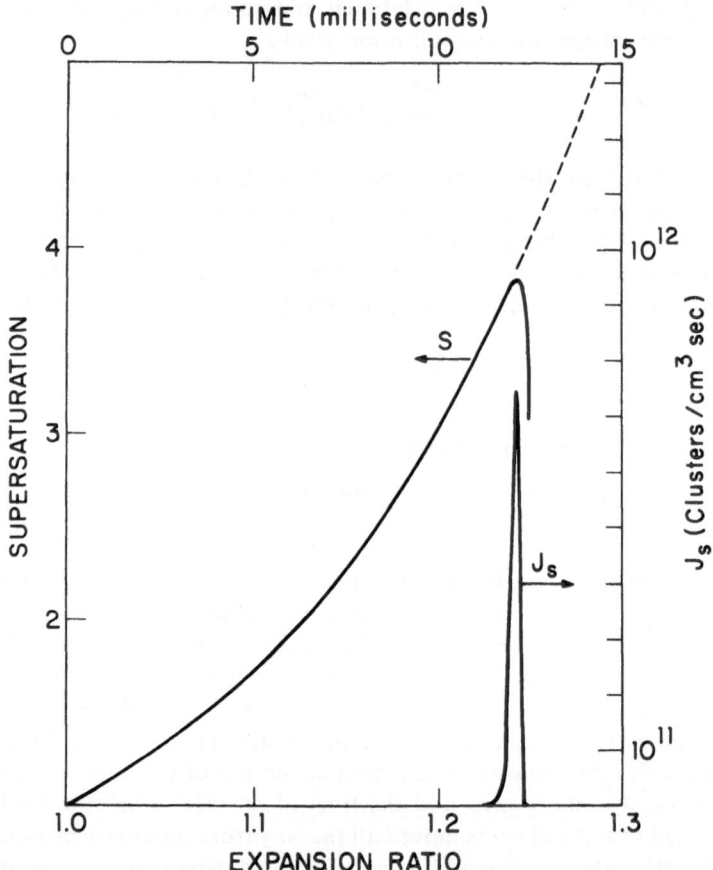

Fig. 4. The actual supersaturation and nucleation rates as functions of expansion ratio or time in a large expansion (2% water vapor in 1 atm air; starting temperature, 293.2°K). Taken from Feder *et al.* [6].

retarding nucleation. Detailed analysis by Feder *et al.* [6] indicates that this effect lowers the flux by a factor typically of the order of 10, but never more than 100—a relatively insignificant correction. τ is increased and $\dot{n}$ is decreased by these same factors.

Clusters nucleated early in an expansion grow fairly rapidly ($\sim 10^{-4}$ sec) to large ($n \sim 10^{10}$ molecules) sizes. These large clusters consume an appreciable fraction of the monomer, thereby reducing the supersaturation and terminating the nucleation process. Figure 4 shows the supersaturation ratio and droplet flux in a chamber the volume of which increases linearly with time. (Expansion ratio is the ratio of the chamber volume to the original volume.) The droplet flux increases very rapidly until the growing clusters suddenly consume the monomer vapor. The width of the peak at half the maximum height is $\sim 10^{-4}$ sec. This was found to be the minimum time for duration of the maximum flux. The maximum duration time is typically 10^{-2} sec.

In comparing theory with experiment, the maximum super-saturation is commonly considered to exist of the order of 1 sec [18]. This establishes the flux as $J = N/t$, where N is the number of droplets nucleated and t is the duration time of the supersaturation, and the flux is compared with that predicted by the various theories. The classical theory [equations (15) or (24)] describes the nucleation of water very well over a temperature range -50 to $100°C$ [1,9,18]. Figure 3 shows the Wilson [19] and Powell [13] data as W–P, and the Volmer–Flood [18] data as V–F. The agreement is seen to be very good. The flux equation calculated from the Lothe–Pound (L–P) $c_0(n)$ is also shown.

Since the number of droplets nucleated is proportional to the sensitive time of the cloud chamber, the interval for which the super-saturation persists, use of the short time of maximum flux increases the calculated flux by a factor of approximately 10^3. Combining this with the factor of $\frac{1}{10}$ in J due to thermal nonaccommodation, one still finds a discrepancy of 10^{14} between the revised theory and experiment.

Not all nucleation data are described by the classical theory, however. Preliminary results on nitrogen [10] are described adequately by the revised theory. Scharrer [16] studied the nucleation on ions of various organic materials, including CCl_4, C_6H_6, C_6H_5Cl, and CH_3Cl. If one allows an approximation for the effect of the ions on the nucleation rate, it appears that the data on these materials are better described by the revised theory.

The experimental situation is clearly unsettled; the nucleation kinetics of water and alcohols are described well by the classical theory, while other materials, including N_2, CCl_4, C_6H_6, C_6H_5Cl, and CH_3Cl, have nucleation kinetics better described by the revised theory. The definitive experiment is yet to be performed and must include measurements on a variety of materials over a range of temperatures under conditions giving strictly homogeneous nucleation.

ACKNOWLEDGMENTS

The kind permission of Professor Philip Hill to cite his unpublished data is gratefully acknowledged. Thanks are due to Mr. Jens Feder, Docent Jens Lothe, and Professor G. M. Pound for permission to cite parts of the work of Feder *et al.* still in press.

REFERENCES

1. A. J. Barnard, *Proc. Roy. Soc. (London)* **A220**: 132 (1953).
2. R. Becker and W. Döring, *Ann. Physik* **24**: 719 (1935).
3. W. J. Dunning, in: *Chemistry of the Solid State*, W. E. Garner (ed.), Butterworths Scientific Publications, Ltd. (London), 1955, p. 159.
4. W. J. Dunning, in: *Abstracts of the Proceedings of International Symposium on Nucleation Phenomena*, Case Institute of Technology (Cleveland), 1965, p. 1.
5. L. Farkas, *Z. Physik. Chem. (Leipzig)* **125**: 236 (1927).
6. J. Feder, K. C. Russell, J. Lothe, and G. M. Pound, *Advan. Phys.* **15**: 1–68 (1966).
7. J. Frenkel, *Kinetic Theory of Liquids*, Oxford University Press (Oxford), 1946.
8. J. W. Gibbs, *The Collected Works of J. W. Gibbs*, Vol. I, Longmans, Green and Co. (New York–London–Toronto), 1878 and 1928.
9. P. G. Hill, in: *Abstracts of the Proceedings of International Symposium on Nucleation Phenomena*, Case Institute of Technology (Cleveland), 1965, p. 16.
10. P. G. Hill, to be published.
11. F. Kuhrt, *Z. Physik* **131**: 185 (1952).
12. J. Lothe and G. M. Pound, *J. Chem. Phys.* **36**: 2080 (1962).
13. C. F. Powell, *Proc. Roy. Soc. (London)* **A119**: 553 (1928).
14. W. H. Rodebush, *Chem. Rev.* **44**: 269 (1949).
15. W. H. Rodebush, *Ind. Eng. Chem.* **44**: 1289 (1952).
16. L. Scharrer, *Ann. Phys. (Leipzig)* **35**: 619 (1939).
17. M. Volmer, *Kinetik der Phasenbildung*, Steinkopff (Dresden–Leipzig), 1939.
18. M. Volmer and H. Flood, *Z. Physik. Chem.* **A170**: 273 (1934).
19. C. T. R. Wilson, *Phil. Trans. Roy. Soc. (London)* **A189**: 265 (1897).
20. J. B. Zeldovich, *Acta Physicochim. URSS* (in English) **18**: 1 (1943).

Nucleation Processes in Deposition onto Substrates

J. P. Hirth and K. L. Moazed

Ohio State University
Columbus, Ohio

CLASSICAL NUCLEATION MODEL

This paper presents a discussion of the following topics: (1) classical development of the theory of heterogeneous nucleation on substrates [1,2]; (2) modifications of the classical theory at the two extremes of low and high substrate temperatures; (3) the effect of substrate imperfections upon nucleation; and (4) the interrelations between nucleation theory and the phenomenon of epitaxy.

Figure 1 illustrates the model that will be adopted in considering the heterogeneous nucleation process. The nucleus is assumed to be a spherical, cap-shaped segment. It is also possible to consider that the critical nucleus is a disk on the surface, but, in general, the kinetics would be much the same [2]; therefore, for simplicity, only the cap-shaped case will be treated here. The amount of the sphere of radius of curvature r intersected on the surface is determined by the equilibrium contact angle θ.

The model of nucleation involves impingement of atoms from the vapor, whether from an uniform vapor phase or from an atomic beam, equilibration on the substrate, surface diffusion, and eventually a surface diffusion jump, which makes the critical-size nucleus grow and become a stable phase on the surface. There are the following two assumptions implicit in this model: (1) that most of the atoms that strike the substrate stick there at least temporarily, or, in other words, that the condensation coefficient is near unity; and (2) that the critical-size nucleus grows by surface diffusion rather than by direct impingement from the vapor. The surface diffusion will be the preferred process if the desorption energy of the adatoms exceeds the activation energy for

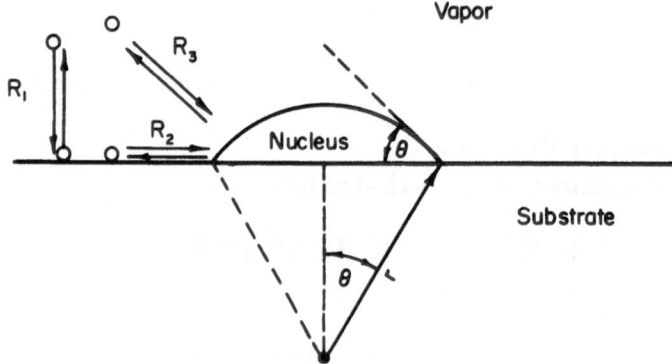

Fig. 1. Model of nucleation involving adsorption from the vapor R_1 and surface
diffusion R_2 to a critical-sized, cap-shaped nucleus.

surface diffusion. Available experimental evidence [2,3] indicates that these approximations are generally valid.

A quantitative development of the ideas expressed above will now be presented. The details of the derivations, which are given in detail elsewhere [1-5], will be omitted here. An attempt will be made, however, to elucidate the physical significance of the formulations. The following equation* represents the concept of equilibration of impinging vapor atoms with adsorbed atoms on the surface:

$$\alpha_c p/(2\pi m k T)^{\frac{1}{2}} = n_1 \nu \exp(-\Delta G_{des}/kT) \tag{1}$$

If equilibrated, the net impingent flux, which is proportional to the vapor pressure p, equals the product of the adatom concentration n_1 and the desorption frequency, given by the product of an atomic vibration term and an exponential involving the desorption energy. If one is not dealing with an uniform vapor phase, but rather with an atomic beam hitting the substrate, one can still consider an effective vapor pressure as expressed by equation (2):

$$p = (J_c/\alpha_c)(2\pi m k T_s)^{\frac{1}{2}} \tag{2}$$

The effective pressure is given by the impingent beam flux J_c and the other factors from the kinetic theory of gases. Provided equilibration prevails, equation (1) indicates that p is directly proportional to n_1; therefore, the supersaturation ratio in the vapor phase is equal to that on the substrate:

$$(p/p_e) = (n_1/n_{1e}) \tag{3}$$

* Notation is defined in a separate section at the end of the paper.

The supersaturation ratio is related to the bulk driving force ΔG_v as follows:*

$$\Delta G_v = -(RT/V)\ln (p/p_e) = -(RT/V)\ln (n_1/n_{1e}) \tag{4}$$

The term ΔG_v is the supersaturation converted to a free energy per unit volume and can be considered to be a driving force for nucleation.

The next few equations break up the free energy of formation of a spherical cap into its component parts. The first part, equation (5), is given by the product of the volume of a spherical cap and the bulk driving force as follows:

$$\Delta G_1 = (4/3)\pi r^3 f(\theta)\,\Delta G_v \tag{5}$$

where $f(\theta)$ is a geometric quantity expressing the fraction of the total sphere intersected by the substrate:

$$f(\theta) = (2 + \cos \theta)(1 - \cos \theta)^2/4 \tag{6}$$

The equilibrium contact angle is a parameter of the substrate and the nucleus and is given by the Young equation as follows:

$$\sigma_{x-v} = \sigma_{c-x} + \sigma \cos \theta \tag{7}$$

The total surface energy of the cap-shaped nucleus is given by equation (8), which, after the substitution of equation (7), can again be expressed in terms of $f(\theta)$;

$$\Delta G_2 = \pi r^2 f(\theta)\sigma + \pi r^2 f(\theta)(\sigma_{c-x} - \sigma_{x-v}) = 4\pi r^2 \sigma f(\theta) \tag{8}$$

The total free energy of formation of one of these spherical, cap-shaped clusters is given by the sum of the volume and surface terms:

$$\Delta G_i = \Delta G_1 + \Delta G_2 \tag{9}$$

To determine the maximum free energy of formation (Fig. 2), the free energy of formation is differentiated with respect to the radius. The maximum size is called the critical size and is denoted by r^*. In carrying out this differentiation, one finds that r^* is given by the Gibbs–Thomson relation, namely,

$$r^* = -2\sigma/\Delta G_v \tag{10}$$

* See P. L. de Bruyn, "Some Aspects of Classical Surface Thermodynamics," this volume, pp. 1–36.

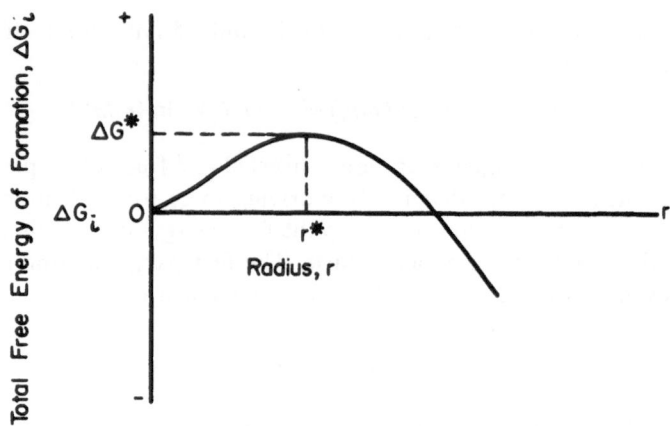

Fig. 2.   Free energy of formation of a nucleus ΔG_i as a function of radius of curvature r.

Substituting equation (10) into equation (9), one finds that the free energy of formation of the critical-sized nucleus is

$$\Delta G_{i^*} = 16\pi\sigma^3 f(\theta)/3\Delta G_v{}^2 \tag{11}$$

It is important to note that ΔG_{i^*} is *inversely* proportional to the square of the driving force (i.e., $\Delta G_v{}^2$); this relationship will be applied later in the discussion of epitaxy and contact angles.

Now given the free energy of formation of clusters of various sizes on the surface, we must consider the minimum free energy for the total surface configuration. In the model shown in Fig. 1, there are single atoms adsorbed on the surface as well as clusters of all sizes up to the critical size. The total free energy of the surface system is given by equation (12) as the summation over all cluster sizes i of the product of the number of clusters of that size and their free energy of formation, minus an entropy-of-mixing term.

$$\Delta G = \sum_i n_i\,\Delta G_i - TS_{\text{mix}} \tag{12}$$

When the adsorbed clusters are localized, the entropy of mixing is given by Fermi–Dirac statistics as

$$S_{\text{mix}} = k \ln \frac{n_0!}{(n_0 - \sum_i n_i)!\,\prod_i n_i!} \tag{13}$$

or

$$S_{\text{mix}} \cong -k \sum_i n_i \ln \frac{n_i}{n_0}$$

where Sterling's approximation is used in the last step. Minimizing the total free energy [equation (12)] with respect to the concentration of critical-sized nuclei,

$$\left(\frac{\partial \Delta G}{\partial n_i} \right)_{r=r^*} = 0 \qquad (14)$$

one finds that the equilibrium concentration of critical-sized nuclei on an equilibrated substrate is given by

$$n_{i^*} = n_0 \exp\left(-\Delta G_{i^*}/kT \right) \qquad (15)$$

Figure 3 illustrates the distribution of cluster concentration as a function of size from monomer ($i = 1$) up to the critical size. The equilibrium curve corresponds to the classical Volmer [6] treatment of nucleation. He assumes that the equilibrium concentration of critical-

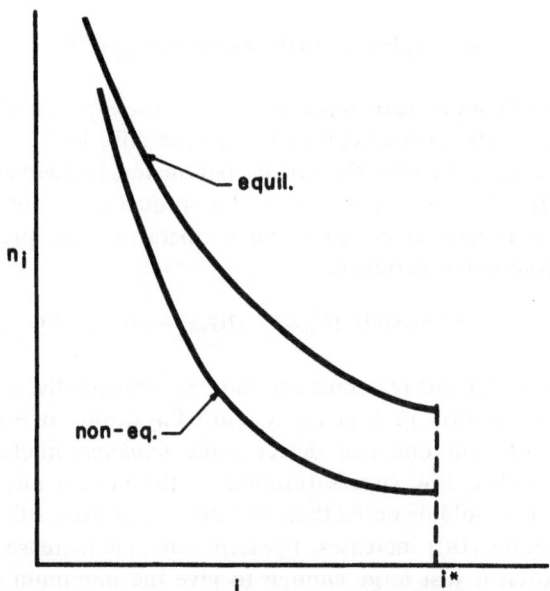

Fig. 3. Equilibrium and quasi-steady-state distributions of clusters as a function of their size i.

size nuclei is maintained and that their rate of growth to a supercritical size is given by a frequency factor. In actuality, the concentration of critical-size nuclei will be depleted by growth to supercritical sizes, and a quasi-steady-state diffusion flux in cluster sizes will be set up. This cluster flux will diminish both the concentration of critical-sized nuclei and the gradient at the critical size.* The solution of this problem for the two-dimensional case [7] indicates that the reduction of both the gradient and the concentration leads to a factor of about 10^{-2} in the nucleation equation; therefore, it is not a critically important factor.

The nucleation-rate equation gives the growth frequency of critical-sized nuclei, which at steady state equals the diffusion flux in cluster sizes, as follows:

$$J = Z\omega n_{i^*} \tag{16}$$

where Z is the factor of $\sim 10^{-2}$ discussed above and the frequency of growth ω is related to the product of the concentration of single adatoms that can join a critical nucleus and their surface diffusion jump frequency, or

$$\omega = n_1 2\pi r^* a(\sin \theta)\nu \exp(-\Delta G_{sd}/kT) \tag{17}$$

The important temperature-dependent term is the exponential involving ΔG_{sd}, the activational free energy for surface diffusion.

Substituting values for the various parameters in these expressions, we can express the nucleation rate as the product of a constant that is relatively insensitive to pressure and temperature, the pressure, and an exponential in temperature:

$$J = C_1 p \exp[(\Delta G_{des} - \Delta G_{sd} - \Delta G_{i^*})/kT] \tag{18}$$

How does one test this resultant equation experimentally, or how does it describe nucleation in a given system? Physically, one sets up an experiment wherein one can detect some *minimum* nucleation rate. One starts with a low supersaturation in the system and gradually increases it. It should be noted that, as $|\Delta G_v|$ increases, ΔG_{i^*} decreases and J [equation (18)] increases. The term $|\Delta G_v|$ is increased until the supersaturation is just large enough to give the minimum observable nucleation rate, and this value is called the critical supersaturation or

---

* See K. C. Russell, "Nucleation in Homogeneous Vapors," this volume, pp. 43–62.

$\Delta G_{v\,\text{crit}}$. In terms of $\Delta G_{v\,\text{crit}}$, equation (18) can be rewritten in the form

$$\left(\frac{1}{\Delta G_{v\text{crit}}}\right)^2 = \left[\frac{3}{16\pi\sigma^3 f(\theta)}\right]\left[kT\left(\ln\frac{C_1}{J} + \ln p\right) + \Delta G_{\text{des}} - \Delta G_{\text{sd}}\right]$$

(19)

where T, J, and p are determined experimentally and values of the parameters in C_1 are generally known. Thus, the test of this nucleation model is to plot $(1/\Delta G_{v\,\text{crit}})^2$ *versus* $kT[\ln(C_1/J) + \ln p]$. If the resultant plot is a straight line, its slope should yield the surface energy function $\sigma^3 f(\theta)$ and its intercept should be $\Delta G_{\text{des}} - \Delta G_{\text{sd}}$. A test of the classical model involves the following questions: Is the plot linear? Do the slope and intercept give reasonable values for the thermodynamic quantities σ, $f(\theta)$, and $\Delta G_{\text{des}} - \Delta G_{\text{sd}}$?

Figure 4 illustrates the above-mentioned type of plot for the deposition of sodium on cesium chloride and on various metal substrates [8]. The plots are linear over the temperature ranges that were considered; and, for the case of sodium on cesium chloride, the contact angle that results from the analysis is 81°, which is a reasonable value. In the case of metal substrates, the contact angle is 101°, which is what

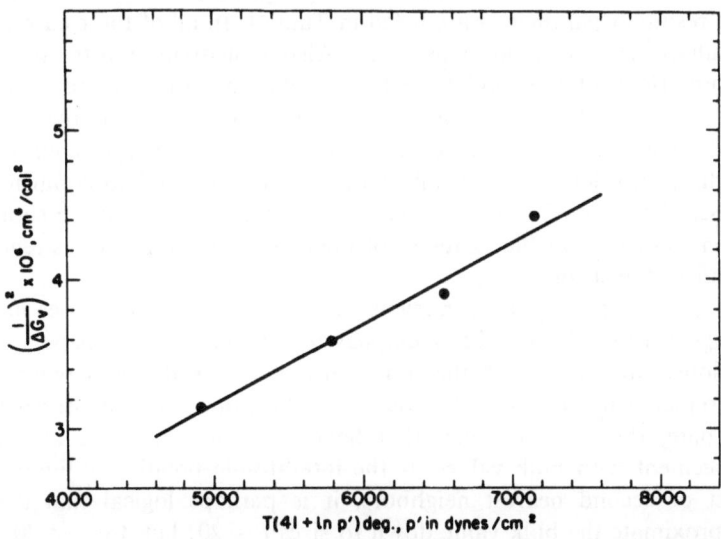

Fig. 4. Plot of equation (19) for the nucleation of sodium on cesium chloride [8].

Table I

Contact Angles and $(\Delta G_{des} - \Delta G_{sd})$ for Various Experiments Fitted to Equation (19)

Deposit	Substrate	θ(deg)	$\Delta G_{des} - \Delta G_{sd}$ (cal/mole)	Reference number
Na	Metal	101	5,750	[8]
Na	CsCl	81	4,220	[8]
Cd	Cu	99	10,300	[9,10]
Zn	Glass	~50	3,110	[11,12]
CrI_2	Alumina	—	26,700	[13]
Cr (from CrI_2 beam)	Alumina	95	23,800	[13]

one might expect for the contact angle of sodium on contaminated metal substrates. These experiments were performed in poor vacua; and this, together with the fact that all the metal results fall on one straight line, further supports the contention that nucleation was occurring on an impurity layer rather than on the metals themselves.

Similar linear plots have been obtained for a number of other systems, the more recent ones involving high vacua of $\sim 10^{-10}$ mm Hg; the resultant parameters are listed in Table I. In all of these cases, the resultant parameters are reasonable. Also, consistent with the original assumptions of the model, sticking coefficients were found to have values $\alpha_c \sim 1$ for cases where the binding energy was appreciable [2] and where the desorption energy exceeded the activation energy for surface diffusion [3]. Thus, all of the above data tend to confirm the classical theory of heterogeneous nucleation. As we shall see shortly, even some data at the extremes of low substrate temperatures tend to confirm the theory.

However, the critical nucleus sizes are small and r^* varies in the range 3–12 Å. As has been emphasized in the previous papers, it is seriously questioned whether one can apply bulk thermodynamics to extremely small clusters of these sizes. Thus, one can only assert that, to date, the contact angles that have been measured are in rough agreement with bulk values. If the interatomic bonding is limited to first or second nearest neighbors, it is perhaps logical that θ will approximate the bulk value down to sizes $i \sim 20$; but, for $i < 20$, the resemblance to the bulk phase should be completely obscured.

Another difficulty with the model is that when data are obtained over a wide substrate-temperature range, curvature appears in the plot of equation (18) [12,13]. This is associated with the onset of different nucleation mechanisms at high temperatures, the topic that is discussed next.

HIGH SUBSTRATE TEMPERATURES

The model that we have assumed so far requires that the adatoms be localized at a given site and that they thermally vibrate in that site until they eventually achieve sufficient thermal energy to be activated and to move to another site, where they are again localized. At high temperatures [high relative to $(\Delta G_{sd}/k)$], this model breaks down and it is more reasonable to describe the adsorbed layer as a two-dimensional gas; the adatoms are no longer associated with a given substrate site. This occurs when the product of the jump distance and the jump frequency is less than the velocity of the atoms in a two-dimensional gas, or

$$a\nu \exp(-\Delta G_{sd}/kT) < (2kT/m)^{\frac{1}{2}} \qquad (20)$$

Insertion of the various atomic parameters essentially reduces equation (20) to the condition $\Delta G_{sd} < kT$. Also, under such conditions, we can conceive of the critical nuclei as both translating and rotating in a two-dimensional gas. Consequently, and analogous to the factors discussed by Russell, one considers contributions to the free energy of formation, which are given by equations (21) and (22) for translation and rotation, respectively;

$$\Delta G_{trans} = -kT \ln(2\pi mikT/n_1 h^2) \qquad (21)$$

$$\Delta G_{rot} = -kT \ln[(8\pi^3 IkT)^{\frac{1}{2}}/h] \qquad (22)$$

Substitution of these factors into the nucleation-rate equation yields

$$J = C_2 p \exp[(\Delta G_{des} - \Delta G_{i^*})/kT] \qquad (23)$$

where the constant C_2, which contains equations (20)–(22), is given by

$$C_2 = Z4\pi^2 r^* kT(\sin\theta)i^*(2\pi IkT)^{\frac{1}{2}}/h^3\nu \qquad (24)$$

Notice that the activation energy for surface diffusion has vanished from this expression. Once again, one can plot the critical super-saturation in the form

$$\left(\frac{1}{\Delta G_{\text{vcrit}}}\right)^2 = \left[\frac{3}{16\pi\sigma^3 f(\theta)}\right]\left[kT\left(\ln\frac{C_2}{J} + \ln p\right) + \Delta G_{\text{des}}\right] \quad (25)$$

Once again, the slope is related to σ and θ, and the intercept in this case is just the desorption energy. Since $C_2 \neq C_1$, the slope of the linear plot should change when the temperature is such that the classical model is modified to the present form. Preliminary results indicate that the curvature in plots over a wide temperature range, for example, that of Fig. 5, can be explained by a transition from equation (19) to equation (25). It should be noted that the condition expressed in equation (20)—$\Delta G_{\text{sd}} < kT$—essentially describes the transition point for high-temperature behavior.

One other modification must be considered at high temperatures and low supersaturation. Because r^* increases wlth decreasing ΔG_v, the concentration of clusters of sizes $i > 1$ may be so large that the mutual impingement of large clusters might be an important nucleation

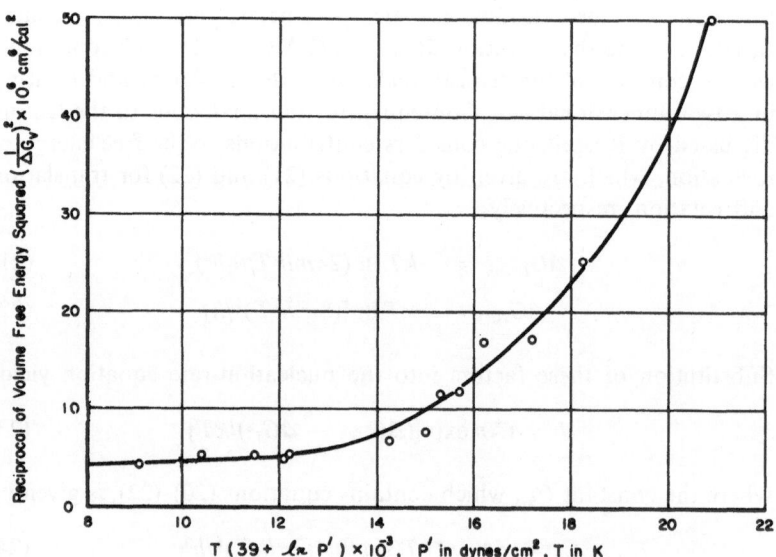

Fig. 5. Plot of equation (19) for the nucleation of CrI_2 on single-crystal alumina [13].

mechanism [5,12]. This process will also be aided by those conditions which would enable the clusters to behave as a two-dimensional gas on the surface of the substrate. For example, the presence of adsorbed gases on the surface of the substrate would increase the rate of diffusion of the clusters on the surface and, hence, increase the frequency of impingement of large-size clusters. There is some evidence for such impingement in nucleus *growth* in the work of Pashley and Stowell [14] and Bassett [15], who observed nucleation and growth directly in transmission electron microscopy.

LOW SUBSTRATE TEMPERATURES

The other limit of applicability of the classical model is low substrate temperatures. Equation (19) can be rewritten in the following form to emphasize its temperature dependence:

$$J = C_3(T) \exp \left[-\frac{16\pi\sigma^3 f(\theta)}{3kT \Delta G_v^2} \right] \tag{26}$$

As the temperature decreases, $C_3(T)$ also decreases. Therefore, in order to maintain a given measurable value of J, the exponential term decreases and ΔG_v *increases*. Thus, r^*, which is inversely proportional to ΔG_v [equation (10)], *decreases* with decreasing substrate temperature. In the limit, the critical nucleus becomes one adatom, and when this occurs, all atoms that hit the substrate stick there and the nucleation model, as such, breaks down. In the range $1 < i^* < 20$, the classical model is severely tested, and a better formal description would involve a statistical-mechanical development for n_{i^*} [16,17]. However, the assignment of partition functions for these small clusters is an exceedingly difficult problem that remains unsolved [4,5]. Thus, at present, one can say only that, somewhere between $i^* = 1$ and $i^* = 20$, the classical description will break down.

The other factor that is important at low substrate temperatures is that the adatom concentration no longer equilibrates with the vapor phase. The factor $\exp(-\Delta G_{des}/kT)$ in equation (1) becomes so small that n_1 can never attain the value n_{1e}. Thus, instead of having a constant adatom population, the adatoms impinging on the substrate continually build up the adatom concentration until a critical supersaturation is achieved *in the adsorbed layer* and nucleation again occurs. The time-dependent adatom concentration is given by

$$n_1(t) = J_c(t) \ll n_{1e} \tag{27}$$

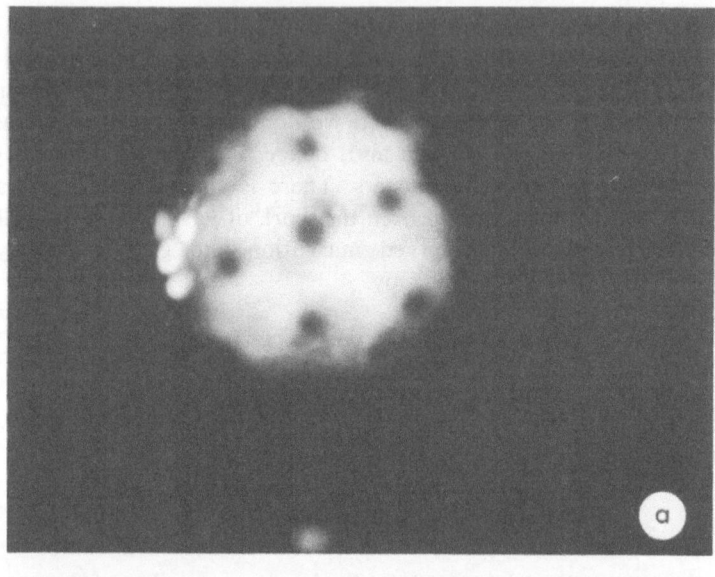

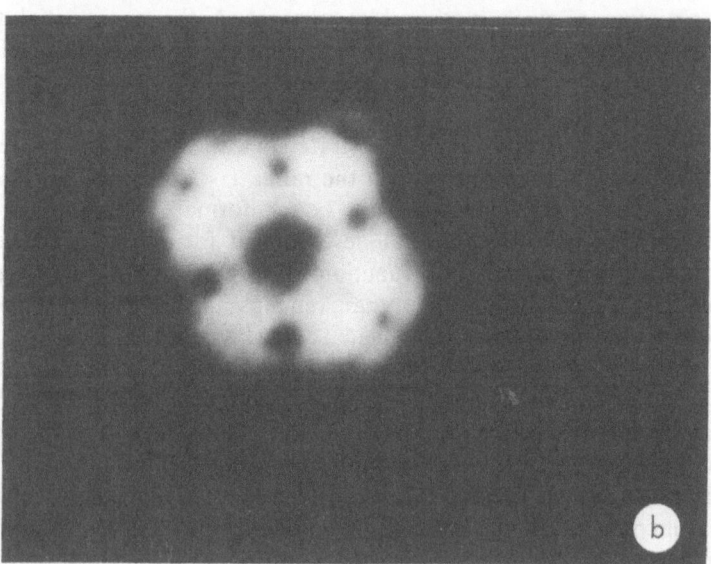

Fig. 6. Field electron emission photographs showing (a) the changes in the emission produced by adsorption and nucleation of silver on tungsten in comparison with (b) a clean tungsten substrate. The very bright spots are silver nuclei.

Experimental studies of nucleation at relatively low temperatures by direct observation in field electron emission microscopes have been performed by Moazed and Pound [18], Gretz and Pound [19], and, quite recently, by Hardy [20]. Figure 6 shows a sequence of micrographs indicating nucleation of silver on a tungsten field emitter tip. In such studies, carried out in ultrahigh vacua, it is possible to work under well-defined, clean-substrate conditions, and, because of the high resolution of the field emission microscope (usually about 20 Å, but in some cases as low as a few angstroms), to observe nucleation from its inception. As indicated in Fig. 6, one can distinguish the adsorbed layer from the surface nuclei on a given surface.

The experiments of Moazed for silver on tungsten and those of Gretz for zinc, cadmium, silver, nickel, and gold on tungsten were carried out at such low temperatures (room temperature to 77°K) that the adatom concentration was time-dependent. Under such conditions, the nucleation model indicates [via equation (27)] that the time for nucleation should be inversely proportional to the impingent flux for a given critical supersaturation. Figure 7 [19] indicates that equation (27) indeed holds in this case. Thus, the *mechanism* of buildup of nuclei in an equilibrated adsorbed layer and eventual growth to supercritical

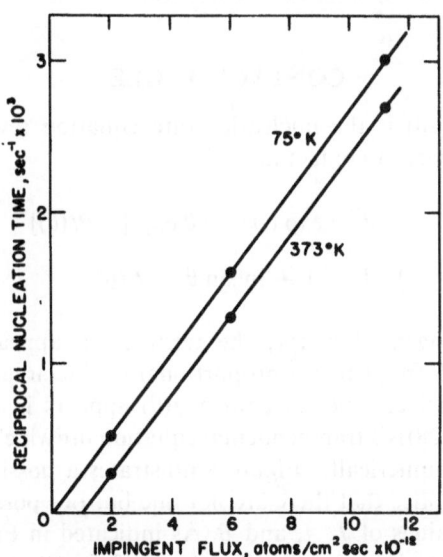

Fig. 7. Test of equation (27) for the case of zinc nucleating on tungsten [19].

size at a critical supersaturation is confirmed even at the low temperature of 77°K, where, as discussed above, it is questionable whether the classical nucleation model should apply; the calculated critical nuclei are in the range 3–10 Å. It is also noteworthy that each nucleus formed a single crystal, consistent with the concept of an embryonic crystallite in equilibrium with monomer.

A troublesome point with respect to the classical theory is that, except for silver (0.3 monolayer), the total adatom concentration corresponded to 1–3 monolayers when nucleation occurred. Experimental observation, however, indicated that a definite adsorbed layer spread over the tip prior to nucleation, and that the nuclei, once formed, were definitely cap-shaped. These results can be rationalized by supposing that a several-atom-thick adsorbed layer forms on the substrate and that "classical" nucleation then occurs on top of this layer. Consistent with this view, desorption-energy measurements for silver [18], mercury [20], and copper [21] from tungsten tips reveal that the first two layers are more tightly bound than subsequent layers. Tiller [22] has noted that two monolayers are required before electronic interactions of new layers with a substrate become small. Thus, the first layers may still appear as a chemically different substrate to the layer in which the cap-shaped nuclei form.

CONTACT ANGLE

Equation (28) is the nucleation-rate equation rewritten to show explicitly the effect of contact angle θ:

$$J = \exp(A)\sin\theta\exp[-Bf(\theta)]$$

$$\ln J = A + \ln\sin\theta - Bf(\theta)$$

(28)

In the pre-exponential factor, the term $\sin\theta$ appears because the diffusion growth frequency is proportional to the circumference of the critical-sized nucleus, and, of course, $f(\theta)$ appears in the exponential term. Equation (28) is a transcendental equation in θ which must be solved graphically or numerically. Figure 8 illustrates a possible solution for the equation; notice that there are not one but *two* possible θ solutions for the given values of J, A, and B. As indicated in Fig. 8, the factor $f(\theta)$ favors nucleation for small values of θ, relating the catalytic potency of the substrate to its wettability by the nucleus. The frequency

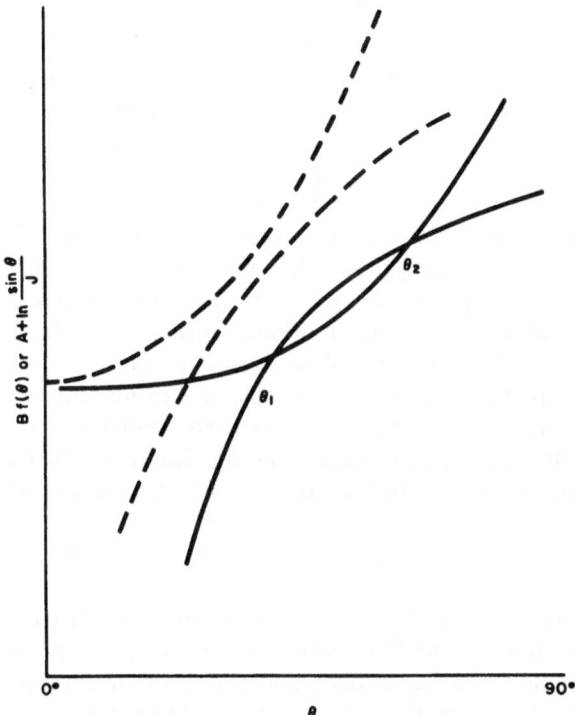

Fig. 8. Plots of $Bf(\theta)$ (bending up) and $A + \ln(\sin\theta/J)$ as functions of the contact angle θ. Dashed curves represent a case in which there is no solution to the transcendental equation (28); solid curves represent a case in which two contact angle values, θ_1 and θ_2, are possible solutions to equation (28).

factor which contains $\sin\theta$, on the other hand, becomes small at θ approaching either 180° or 0°, because the circumference of the critical nucleus becomes small where it contacts the sphere. In the limit of θ approaching zero, of course, the fact that equation (28) does not have a solution has no physical significance in all likelihood. Rather, it indicates that a mechanism other than the one discussed here should become applicable in such a limit [23]. It is important, however, to recognize that critical conditions can be satisfied by two contact angles, e.g., 15° and 65°, both of which are physically possible. A choice between the two must be based on considerations other than the nucleation experiment itself [23].

SUBSTRATE IMPERFECTIONS

Up to this point, we have been considering a cap-shaped nucleus on an idealized flat substrate. Imperfections in the substrate will, in general, affect the nucleation process. Suppose that there are monatomic steps on the surface associated with screw dislocations, growth steps, *etc.* As indicated in Fig. 9, an additional little area of substrate–vapor interface is removed and replaced by substrate–condensate interface if the nucleus forms at the step instead of on the flat. The latter surface has a lower surface energy, and thus the free energy of formation of the nucleus at a step will be lower than one upon the flat. In essence, any configuration in which additional surface area is removed in forming the critical nucleus will favor nucleation.

Figure 10 shows another favorable nucleation site, a re-entrant groove on the substrate. This case has been treated quantitatively [24] and illustrates the specific effects on the frequency factor and the driving-force term. The free energy of formation is less at the step, favoring nucleation there; but the concentration of imperfection sites is usually less than the concentration of sites on the flat. Because the nucleation current I is the product of the nucleation rate and the concentration of available sites per unit area, the latter factor will favor nucleation on the flat surface. Figure 11 presents the $f(\theta)$ and $K(\theta)$ functions for a flat surface and for a 90° re-entrant angle [24], respectively. The nucleation rate at a step will be given by equation (16)

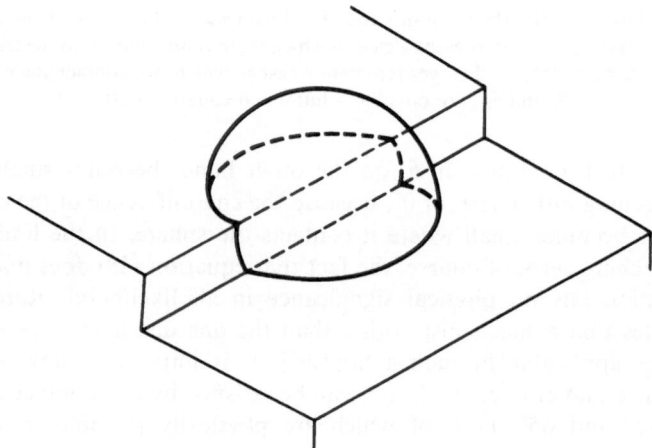

Fig. 9. Nucleation at a surface step.

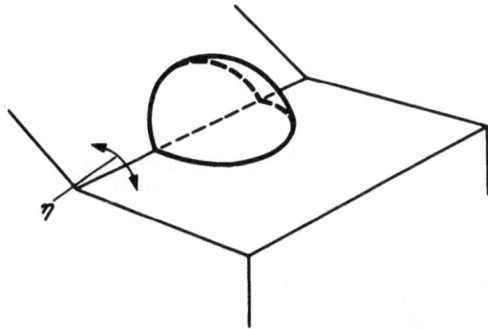

Fig. 10. Nucleation at a re-entrant groove.

with $K(\theta)$ replacing $f(\theta)$ in ΔG_{i^*} [equation (11)]. For a given value of θ, $K(\theta)$ is always less than $f(\theta)$; therefore, as was stated earlier, the free energy of formation is less at the groove. Figure 12 is a plot of the

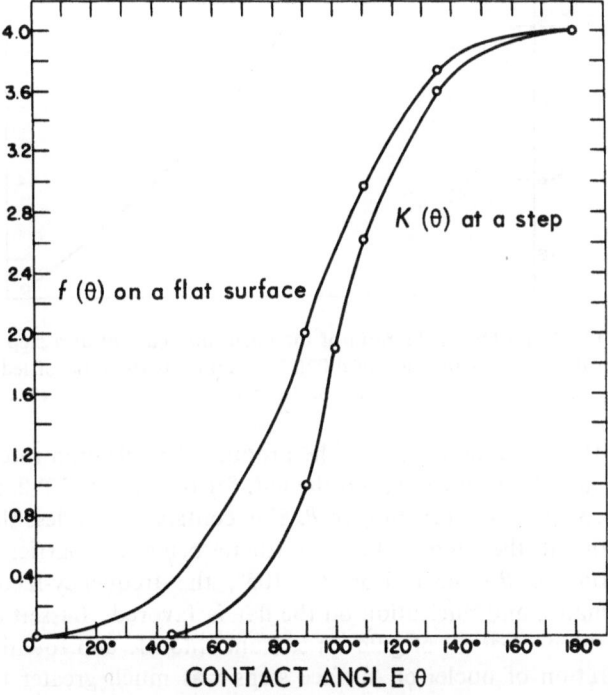

Fig. 11. Geometric factors $f(\theta)$ for nucleation on a flat surface and $K(\theta)$ for nucleation at a step as functions of θ.

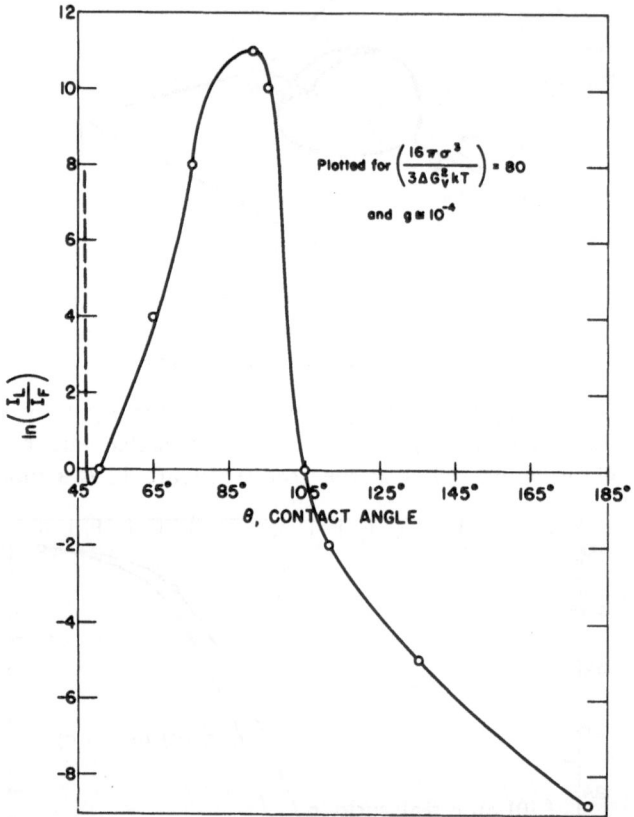

Fig. 12. The logarithm of the ratio of the nucleation current at a step I_L to that on a flat substrate as a function of θ [24]. The term g is the ratio of ledge sites to sites on the flat.

ratio of the nucleation current (the product of nucleation rate and site density) at a 90° step to that on the flat, for the case of 10^4 flat sites for every step site, as a function of θ. For contact angles less than 105°, nucleation at the step is favored (there being no barrier to such nucleation for $\theta < 45°$). For $\theta > 105°$, the frequency-factor effect predominates, and nucleation on the flat is favored. Bassett et al. [25] studied the nucleation of gold on sodium chloride and found that the concentration of nuclei on surface steps was much greater than that on the flats. Since the contact angle is 90° in this case, the findings are consistent with theory [24,25]; the nucleation rate on steps should be greater, but since $90°$ is close to the critical angle of $105°$, some nucleation on the flats should be observed. In a similar study of cadmium deposition on lithium fluoride [24], nucleation was found to occur preferentially on the flats, which is consistent with the finding that the contact angle is $120°$ in this case.

The above development was for the special case of a $90°$ groove. In general, every substrate imperfection requires a specific treatment, but qualitatively the effects will be similar. All surface imperfections will tend to be preferential nucleation sites for cases where the contact angle is small. In addition, there is a possibility of impurity adsorption on the substrate. Without going into detail on this point, it should be noted that the principal effect of impurities is to change the various surface energies via surface adsorption [2].

EPITAXY

The final point that is considered here concerns the interrelation of nucleation theory and epitaxy. Reconsider equation (28), which expresses the nucleation rate as a function of contact angle θ. Consider now the ratio of the nucleation rate for a favorable epitaxial orientation, which can be characterized by a contact angle θ_1, and for an unfavorable random orientation, characterized by θ_2. All terms in equation (28) except those involving θ will be the same in both cases; therefore, the ratio of the rates of epitaxial to random nucleation will be

$$\beta = \frac{J_1}{J_2} = \frac{\sin \theta_1}{\sin \theta_2} \exp \{B[f(\theta_2) - f(\theta_1)]\} \tag{29}$$

Now $\theta_2 > \theta_1$; thus, $f(\theta_2) > f(\theta_1)$. Furthermore, as was discussed in relation to equation (26), the pre-exponential factor increases with increasing temperature, so that B also increases with increasing substrate temperature. Therefore, high substrate temperatures produce a large B; hence, a high ratio β, and, hence, favor epitaxial nucleation. Also, high substrate temperatures, or large B, are associated with *low* supersaturations [equation (26)] and, thus, with relatively *large* critical nucleus sizes [equation (10)]. This is an important point because the other theoretical development for epitaxy [26] requires epitaxy to be the result of the orientation of small (2- or 3-atom) critical nuclei. However, these small nuclei can occur only at *low* substrate temperatures, and at these temperatures equation (29) predicts that random nucleation should be about as prevalent as favorable nucleation; thus, epitaxy should not occur.

Finally, it is noted that either of the conditions $\theta \to 0°$ or $\theta \to 180°$ leads to a small value of the pre-exponential factor, a small value of B, a small value of $[f(\theta_2) - f(\theta_1)]$, and, hence, a small β. *Intermediate* values of the contact angle ($\theta \sim 50°$) are the most favorable for epitaxial deposition. Thus, relatively high substrate temperatures, small driving forces, and intermediate values of the contact angle are required for epitaxy. There is some experimental evidence [27] that supports this view, but further work is required to establish it unequivocally.

ACKNOWLEDGMENT

This work was supported by the U. S. Office of Naval Research.

NOTATION

a = Diffusion jump distance

ΔG_{des}, ΔG_{sd} = Activational free energies for desorption and surface diffusion, respectively

ΔG_i, ΔG_{i^*} = Free energies of formation of ith-size cluster and the critical-sized cluster, respectively

ΔG_v = Bulk free energy change per unit volume

ΔG_{trans}, ΔG_{rot} = Translational and rotational contributions to the free energy of formation of a cluster, respectively

h = Planck constant

i = Number of atoms in a cluster

I = Moment of inertia of surface cluster

J = Nucleation rate

J_c = Vapor impingement flux

k = Boltzmann constant

m = Atomic mass

n_0 = Concentration of substrate sites

n_1, n_{1e} = Concentration and equilibrium concentration, respectively, of adsorbed monomer

n_i, n_{i^*} = Concentration of clusters and of critical-sized clusters, respectively

p, p_e = Effective vapor pressure and equilibrium vapor pressure, respectively

R = Gas constant

r^* = Radius of curvature of critical nucleus

T = Absolute temperature

T_s = Substrate temperature

V = Molar volume

Z = Nonequilibrium (Zeldovitch) correction factor

α_c = Sticking coefficient for vapor atoms striking the substrate

β = Ratio of rates of epitaxial to random nucleation.
θ = Contact angle
ν = Atomic vibrational frequency
$\sigma, \sigma_{x-v}, \sigma_{c-x}$ = Surface energies of the nucleus–vapor, substrate–vapor, and nucleus–substrate interfaces, respectively
ω = Frequency of growth of a critical-sized nucleus

REFERENCES

1. G. M. Pound, M. T. Simnad, and L. Yang, *J. Chem. Phys.* **22**: 1215 (1954).
2. J. P. Hirth and G. M. Pound, *Condensation and Evaporation*, Pergamon Press, Inc. (Oxford), 1964, pp. 41–76.
3. G. Ehrlich, in: *Structure and Properties of Thin Films*, C. A. Neugebauer, J. B. Newkirk, and D. A. Vermilyea (eds.), John Wiley & Sons (New York), 1959, p. 423.
4. G. M. Pound and J. P. Hirth, in: *Condensation and Evaporation of Solids*, E. Rutner, P. Goldfinger, and J. P. Hirth (eds.), Gordon & Breach (New York), 1964, p. 475.
5. J. P. Hirth, S. J. Hruska, and G. M. Pound, in: *Single Crystal Films*, M. H. Francombe and H. Sato (eds.), Pergamon Press, Inc. (Oxford), 1964, p. 9.
6. M. Volmer, *Kinetik der Phasenbildung*, Steinkopff (Dresden), 1939.
7. J. P. Hirth, *Acta Met.* **7**: 755 (1959).
8. L. Yang, C. E. Birchenall, G. M. Pound, and M. T. Simnad, *Acta Met.* **2**: 462 (1954).
9. J. D. Cockcroft, *Proc. Roy. Soc.* (*London*) **119A**: 293 (1928).
10. S. J. Hruska and G. M. Pound, *Trans. Met. Soc. AIME* **230**: 1406 (1964).
11. J. B. Hudson, *J. Chem. Phys.* **36**: 887 (1962).
12. S. J. Hruska, *Acta Met.* **12**: 1211 (1964).
13. C. S. Jackson, R. D. Gretz, and J. P. Hirth, to be published.
14. D. W. Pashley and M. J. Stowell, *Phil. Mag.* **8**: 1605 (1963).
15. G. A. Bassett, in: *Condensation and Evaporation of Solids*, E. Rutner, P. Goldfinger, and J. P. Hirth (eds.), Gordon & Breach (New York), 1964, p. 599.
16. J. P. Hirth, *Ann. N. Y. Acad. Sci.* **101**: 805 (1963).
17. T. N. Rhodin and D. Walton, *Metal Surfaces*, ASM (Cleveland), 1963, p. 259.
18. K. L. Moazed and G. M. Pound, *Trans. Met. Soc. AIME* **230**: 234 (1964); K. L. Moazed, Ph.D. Thesis, Carnegie Institute of Technology, Pittsburgh, Pennsylvania, 1959.
19. R. D. Gretz and G. M. Pound, in: *Condensation and Evaporation of Solids*, E. Rutner, P. Goldfinger, and J. P. Hirth (eds.), Gordon & Breach (New York), 1964, p. 575; R. D. Gretz, Ph.D. Thesis, Carnegie Institute of Technology, Pittsburgh, Pennsylvania, 1963, to be published.
20. S. C. Hardy, *International Symposium on Nucleation*, Case Institute of Technology, Cleveland, Ohio, April 1965.
21. J. S. Anderson and J. P. Jones, *10th Field Emission Symposium*, Berea, Ohio, 1963.
22. W. A. Tiller, *International Symposium on Nucleation*, Case Institute of Technology, Cleveland, Ohio, April 1964.

23. K. L. Moazed and J. P. Hirth, *Surface Sci.* **3**: 49 (1965).
24. B. K. Chakraverty and G. M. Pound, *Acta Met.* **12**: 851 (1964); B. K. Chakraverty and G. M. Pound, in: *Condensation and Evaporation of Solids*, E. Rutner, P. Goldfinger, and J. P. Hirth (eds.), Gordon & Breach (New York), 1964, p. 553.
25. G. A. Bassett, J. W. Menter, and D. W. Pashley, *J. Inst. Metals* **26**: 449 (1959); G. A. Bassett *et al.*, in: *Structure and Properties of Thin Films*, C. A. Neugebauer, J. B. Newkirk, and D. A. Vermilyea (eds.), John Wiley & Sons (New York), 1959, p. 12.
26. D. Walton, *J. Chem. Phys.* **37**: 1282 (1962); *Phil. Mag.* **7**: 1671 (1962).
27. B. W. Sloope and C. O. Tiller, *J. Appl. Phys.* **32**: 1331 (1961); **33**: 3461 (1962).

Nucleation and Condensation in Polymer Systems

Fraser P. Price

General Electric Research Laboratory
Schenectady, New York

INTRODUCTION

Synthetic high-polymeric materials, particularly crystallizable ones, have the following two attributes which sharply distinguish them from most other substances: (1) essentially all of them were unknown as recently as thirty years ago, and (2) their molecules have an essentially thread-like shape. Liquids of metals or low-molecular-weight molecules can, to a good approximation, be regarded as aggregates of spheres, or at worst, ellipsoids of small axial ratio. On the other hand, the length of a typical, synthetic, high-polymer molecule is several thousand times its width. Thus, it is not surprising that synthetic high polymers exhibit properties that are very different from those of low-molecular-weight substances, e.g., rubbery behavior. The molecules of synthetic high polymers, being held together by covalent bonds, usually remain inviolate throughout the manipulations to which they are subjected. This is in contradistinction to the melting behavior of inorganic glasses where, because of structure, there must be interchange of bonds between neighboring atoms [1].

A polymer molecule, either in a pure polymer melt or in solution, pervades a volume some five to twenty times the volume actually occupied by the backbone of the molecule itself [2]. Thus, in a molten polymer, the volume in which some part of a polymer molecule can be found contains of the order of ten other polymer molecules. This means that there is considerable interpenetration of these thread-like molecules, with concomitant chances for entanglement. Surprisingly, in spite of this entangled situation, there are many polymers which (if put into an appropriate range of thermodynamic parameters) can crystallize [3]. This is evidenced by abrupt changes in such properties as density and heat content and, most particularly, by development of

X-ray diffraction patterns characteristic of three-dimensional crystalline order.

An attempt is made here to survey, in quite broad terms, the nature of the crystals developed and the mechanisms by which they are produced. It is not intended that this survey be exhaustive. Rather, it is hoped that this discussion will afford the interested reader a panoramic view of the field at this moment in time and also will serve the more serious student as a starting point for further study.

CRYSTALLIZATION FROM SOLUTION

Morphology

When crystallization occurs in bulk, because of the intertwining of the molecules, it is impossible to separate out single crystals for observation and study. It was not until 1957 that isolated single crystals of synthetic high polymers were prepared and described. In that year, reports from three laboratories [4-6] described the preparation of crystals of linear polyethylene by slow cooling of a very-low-concentration (0.01%) solution of this polymer in p-xylene. In retrospect, the use of the extreme dilution was clearly necessary to minimize interpenetration of the molecules. The crystals formed were very small,

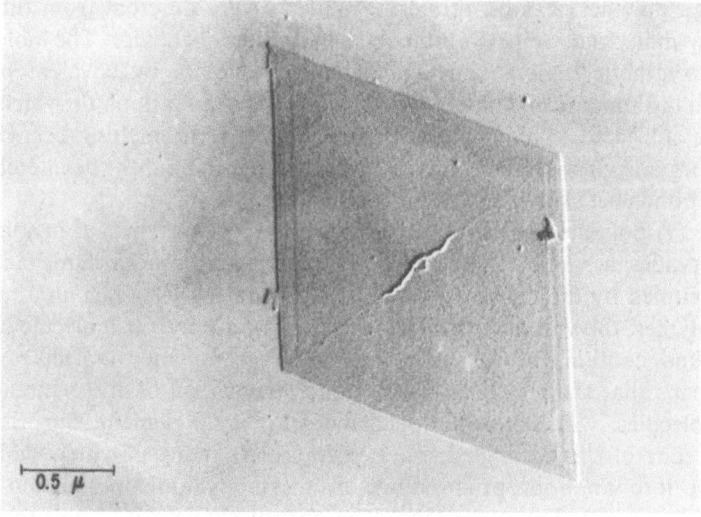

Fig. 1. Electron micrograph of a polyethylene single crystal.

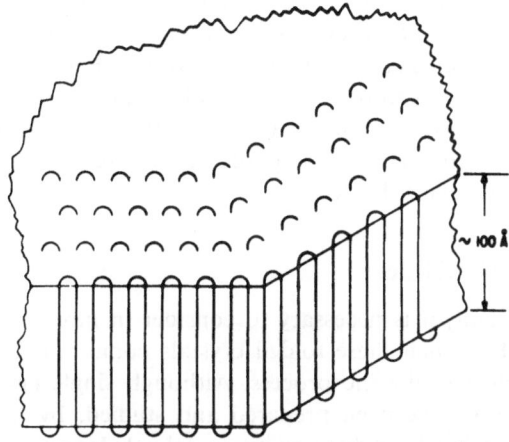

Fig. 2. Schematic diagram of the edge of a polymer single crystal.

flat plates that could be observed only by use of the electron microscope. One such crystal, prepared in the author's laboratory, is shown in Fig. 1. It is seen that this is a diamond-shaped plate the large dimensions of which measure a few microns. The height of the plate, as judged by the width of the shadow, is about 100 Å. Selected area electron diffraction patterns of such crystals indicate that the C-axis (the chain axis) is perpendicular to the large, flat faces. Thus, the polymer chains are oriented normal to the large, flat faces. Since the polymer chains in polyethylene average 10,000 Å in length and since the crystal is only about 100 Å thick, if one demands that the chain be completely crystallized, it must fold back and forth and traverse the crystal many times. This folding is represented schematically in Fig. 2. This diagram makes it clear that the top and bottom surfaces of these crystals are planes of folds. There has been considerable discussion over the last several years as to whether or not these folds are all identical and occur in four- or five-carbon atoms, or whether or not they have a rather broad statistical distribution of lengths and shapes. Without going into particulars, most knowledgeable investigators of the phenomena now agree that, in crystals from dilute solution, the folds are sharp, regular, and identical. The electron diffraction patterns and the angles between the vertical faces of the crystal in Fig. 1 indicate that the vertical faces are the various [110] faces derivable from successive twinning across [100] and [010] planes [7]. These vertical faces are the growth faces of the crystal. The crystal shown in Fig. 1, which is the simplest of a wide range of forms exhibited by polyethylene, seems to be a flat plate; but closer investigation has shown that, when it comes from solution, it is, in reality, a pyramid [7]. It is not a simple flat-based pyramid, however; for further discussion of this and other aspects of the crystals morphologies, the reader is referred to a recent excellent review by Keller [8].

Mechanism and Kinetics

At this point, it is necessary to consider in more detail both the mechanisms by which these folded crystals come into being and the kinetics which describe the process. Although single crystals of over fifteen polymers have been prepared and studied, by far the largest amount of work has been done on linear polyethylene precipitated from dilute solution. For this reason, the subsequent discussion of kinetics of crystallization from dilute solution will be restricted to polyethylene.

Soon after the discovery of the techniques for single-crystal preparation, it was found that: (1) upon isothermal precipitation, the thickness of the crystals increased with increasing precipitation temperature; (2) if the temperature was raised or lowered stepwise, the thickness of a given crystal likewise increased or decreased; and (3) with isothermal preparation, the thickness of a given crystal was very constant and, furthermore, all the crystals had the same thickness. For several years there was considerable speculation concerning the growth mode of these polymer crystals, and, finally, in 1959, it occurred independently and essentially simultaneously to several investigators that the four observations listed above could be explained in terms of a process of nucleation-controlled growth [9,10]. Actually, this was a resurrection of the ideas of Volmer [11], who had proposed some forty

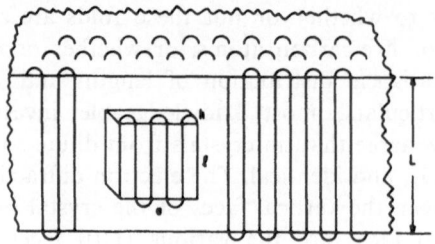

Fig. 3. Model for growth nucleus.

years earlier that the rate-controlling step in the growth of crystals was the nucleation of tiny new patches of crystals upon and coherently with an already completed crystal face. The model for the growth of polymer crystals is shown in Fig. 3. Here is shown a completed growth face of height L to which adheres a coherent patch of new crystal of dimensions a, h, l. First, an inquiry about the free energy required to produce such a patch shows that this free energy can be expressed as follows:

$$\Delta g = -ahl\,\Delta G_v + 2aho_e + 2lho_s \tag{1}$$

where ΔG_v is the free energy of solution per unit volume at the temperature under consideration, σ_e is the surface energy of the ah face, and σ_s is the surface energy of the lh face. It can be shown that equation (1) represents a saddle-shaped surface if σ_e, σ_s, and ΔG_v are fixed. Determination of the values of Δg, a, and l at the saddle point is accomplished by letting $\partial(\Delta g)/\partial a = 0$ and by letting $\partial(\Delta g)/\partial l = 0$. This leads to the following particular values of l, a, and Δg:

$$l^* = (2\sigma_e)/\Delta G_v$$

$$a^* = (2\sigma_s)/\Delta G_v$$

and

$$\Delta g^* = (4h\sigma_s\sigma_e)/\Delta G_v$$

Furthermore, if it is assumed that the heat and entropy of solution do not vary significantly from the solution temperature down to the temperature of interest,

$$\Delta G_v = \Delta S_v(T_m - T)$$

where ΔS_v is the entropy of solution at the solution temperature T_m and where T is the temperature of interest. Thus, one can write

$$\Delta g^* = \frac{4h\sigma_s\sigma_e}{(\Delta S_v)(\Delta T)} \tag{2}$$

where $\Delta T = T_m - T$.

Now assume that once a nucleus forms with a certain length l growth occurs only in the a direction (the length remains constant). This is not unreasonable; for, if the nucleus is longer than the substrate length L, its formation becomes increasingly difficult; and, if the nucleus becomes too short (less than l^*), the energy of the folds becomes overwhelming and the nucleus becomes unstable. Thus, there is a balance between energies, which tends to limit the length. In this situation, it is clear that the length l (the thickness) will vary with temperature in the observed manner and will be sensitive to both increasing and decreasing temperature. Since the length l of an adding layer is assumed fixed once the nucleus length is fixed, the distribution of lengths both within and among crystals is determined by the distribution of nucleus lengths. It can be readily shown that this distribution of nucleus lengths is within the observed limits of crystal roughness and thickness [9,10]. For reasonable values of the parameters for polyethylene, 90% of the nucleus lengths are within 10% of the average. Now, if the rate of growth is controlled by the production of nuclei of sizes in the range of a^*, l^*, it can be shown that this growth rate is

$$G = G_0 e^{-E^{\pm}/kT} e^{-\Delta g^*/kT} \qquad (3)$$

where G_0 is a temperature-independent constant involving segmented properties of the molecule, $E^{\pm}$ is an activation energy involved in transport of material across the liquid–crystal boundary, and k is Boltzmann's constant. Equation (3) implies that, for nucleation-

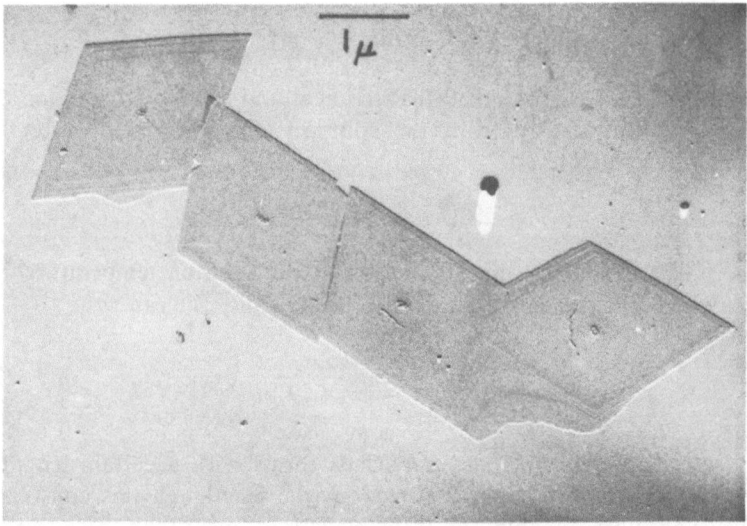

Fig. 4. Single crystals of polyethylene showing evidence of homogeneous primary nucleation.

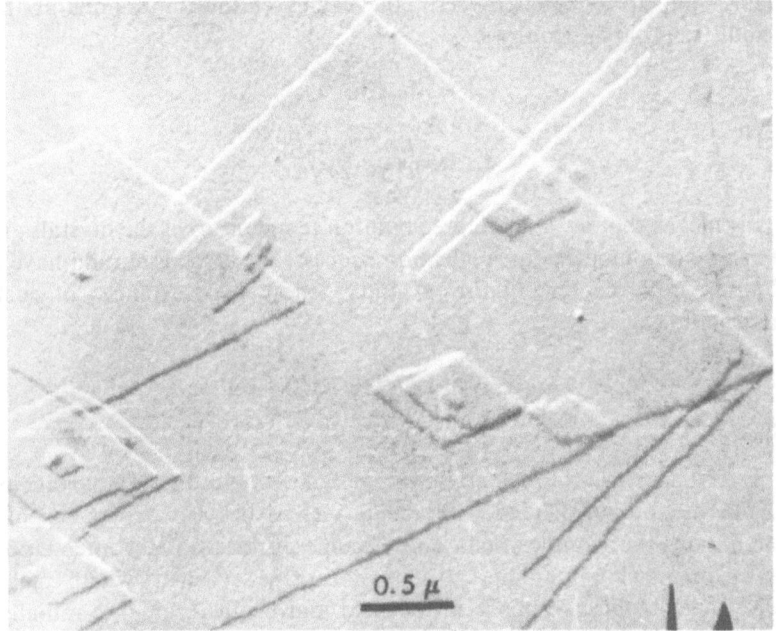

Fig. 5. Single crystals of poly-3,3-bischloromethyloxacyclobutane showing evidence of homogeneous primary nucleation. [Courtesy P. H. Geil, *Polymer* 4: 404 (1963).]

controlled growth at low supercooling, plots of log G *versus* $1/T\Delta T$ should be straight lines. One set of experiments [12] on growth rates of polyethylene crystals is consistent with this conclusion, but the data are not sufficiently precise to distinguish between this and other possible growth modes. Equation (3) will be considered again when the kinetics of bulk crystallization are discussed.

Thus far, only the growth processes have been considered. This has been done because it is the growth modes that determine the morphology. However, there is one instance where the birth, that is, primary nucleation, of the crystals is worth considering. However, it is necessary to consider first the energetics of the process of production of a small crystal all by itself in contact with the solution from which it came. This leads to an expression similar to equation (1) with one added term ($2al\sigma_s$) that takes into account the two *al* faces which here are in contact with the solution. If the partial differentials of such an expression are equated to zero, the values of the dimensions at the saddle point are as follows:

$$l_p{}^* = 4\sigma_e/\Delta G_v$$

and

$$a^* = h^* = 2\sigma_e/\Delta G_v$$

This means that at a given precipitation temperature if the crystals are born by a homogeneous nucleation process, the crystals should have a bump in the center. Figures 4 and 5 show two instances of such morphology.

CRYSTALLIZATION IN BULK

Morphology

If a specimen of a crystallizable polymer is put on the hot stage of a microscope and heated while being viewed between crossed nicols, eventually the sample melts and becomes optically isotropic. Upon subsequent slow cooling, a temperature is reached where small birefringent objects appear in the field and begin to enlarge radially. After a short time, the field might appear as shown in Fig. 6, which displays the spherically symmetric, birefringent objects (spherulites) which develop in nylon 66. If the growth of these spherulites is permitted to go to completion, the whole field becomes filled with the polygonized structures resulting from their abutment. These spherulites are not single crystals, but rather are organized arrays of single crystals. Furthermore, it appears from microbeam X-ray experiments that all the crystals are contained within the spherulites. The black cross in each spherulite results from its behaving like an array of radiating anisotropic needles. Such an array when viewed between crossed nicols shows extinction in the four positions where the optic axes of the needles are parallel to the nicol directions.

Spherulites are classified as positive or negative depending upon whether the radial refractive index exceeds the tangential one or *vice versa*. Most polymer spherulites are negative, a situation which results from the polymer chains always lying in an essentially tangential position [13]. The nylons are different, however, being negative if formed at temperatures only slightly below the melting point and positive if formed at lower temperatures [14-17]. The spherulites shown in Fig. 6 were made by cooling a melt of nylon 66 from above its

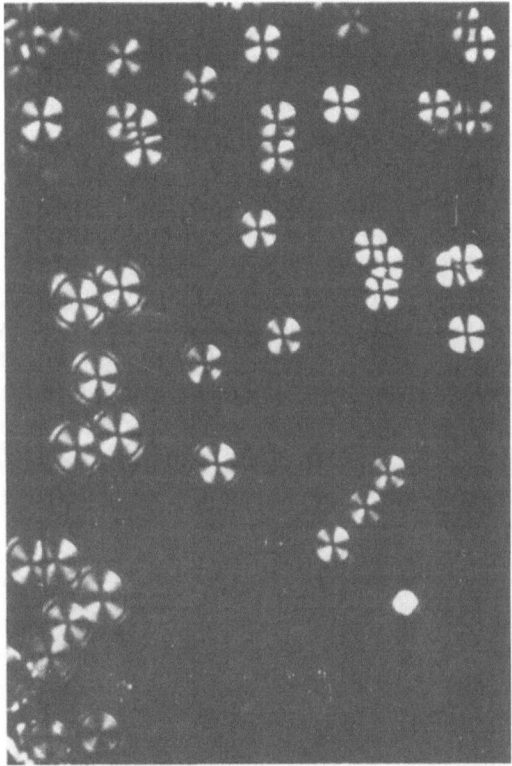

Fig. 6. Spherulites of nylon 66 developing from the melt.

melting temperature. Thus, the birefringence of the centers of the spherulites is negative, while the birefringence of the outer portions is positive. Separating these two regions is a circular black band of ostensibly zero birefringence. The change in sign of the birefringence is dramatically illustrated by the color plate (facing p. 94), which shows several frames from a color motion picture where the background has filled in with many smaller spherulites. The color is largely due to the insertion of a first-order, red, gypsum plate in the optical train of the microscope. This produces compensation leading to alternating yellow and blue quadrants as the spherulite is examined at fixed radius. In this color plate, the quadrants that are blue in the center are yellow in the outer regions. This unequivocally shows that the sign of birefringence changes in going across the black band.

These are the objects, the polycrystalline arrays, that are observed in the crystallization of polymers in bulk; and it is of some significance to inquire as to whether a not the plate-like morphology observed in dilute-solution crystallization has any bearing on bulk-crystallized materials. In recent years, much evidence has been developed which indicates that the substructure of the spherulites consists of thin lamellae [18]. Figures 7 and 8 show electron micrographs of shadowed replicas of spherulites which developed on a free surface of molten, supercooled, isotactic polystyrene. These figures, which show the two types of objects formed, both exhibit a lamellar substructure. Figure 7 appears to be an array of platelets observed face-on. Furthermore, the whole array has a hexagonal symmetry. Figure 8 also shows a lamellar structure which, in the center, at least, appears to be standing on edge. Both these

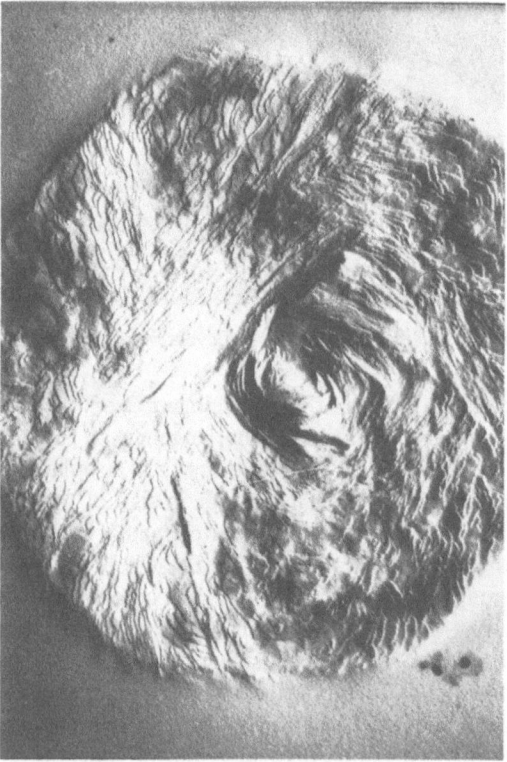

Fig. 7. Electron micrograph of shadowed-replica spherulite of isotactic polystyrene.

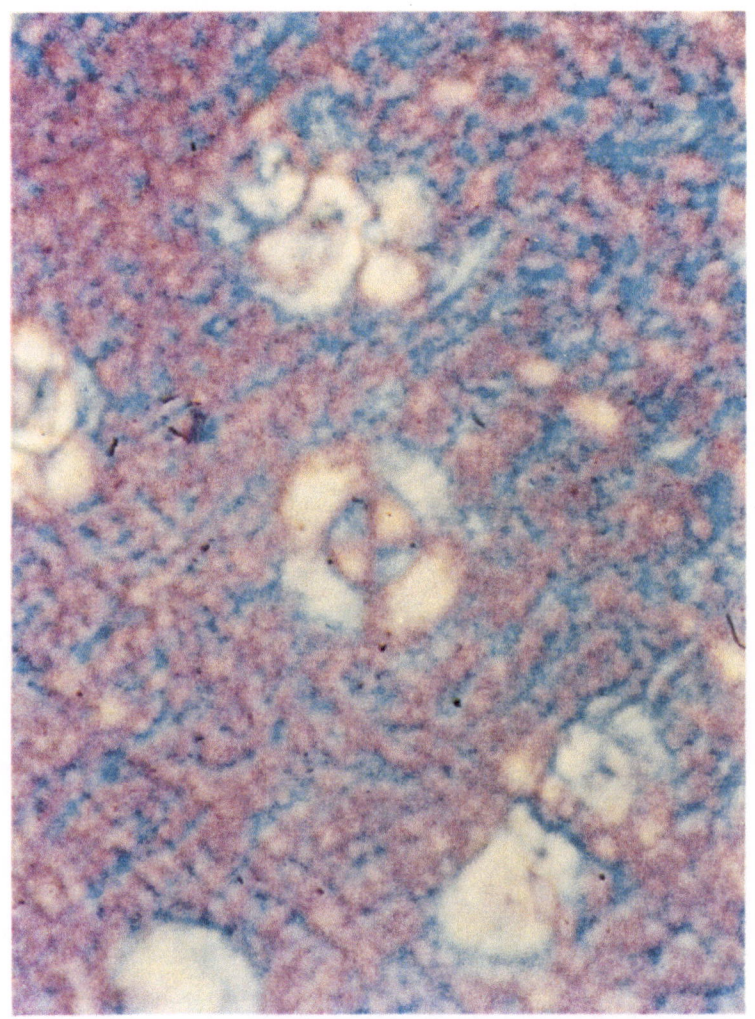

Spherulites developing in nylon.

Fig. 8. Electron micrograph of shadowed-replica spherulite of isotactic polystyrene.

structures could result from a screw-dislocation mode of growth. In this event, Fig. 7 shows the objects viewed down the screw axis, while Fig. 8 shows them viewed perpendicular to this axis. These objects, called hedrites by some, have been observed in the polarized-light microscope [19]; their optical behavior is consistent with the screw-dislocation growth mode.

The electron micrographs of Figs. 7 and 8 are of the surfaces of crystalline polymers, and it is of some pertinence to inquire as to whether or not the lamellar habit persists inside bulk-crystallized samples. Recently, it has been found that fuming nitric acid will disintegrate crystalline polyethylene into fragments suitable for electron microscopic study [20]. An electron micrograph of some of the debris

from such a treatment [21] is shown in Fig. 9. Here it is quite obvious
that the objects thus found are plate-like in habit. Space does not
permit further exposition of this work and, for further study, the
reader is referred to the recent literature [21,22]. It is sufficient to note
that it appears that lamellae characterize the internal as well as the
surface structure of bulk-crystallized polymers. This does not mean,
however, that the lamellae in bulk are completely of the chain-folded
type, nor are the folds as perfect as in solution-precipitated crystals.
In fact, aside from the difficulty of visualizing how such perfect,
regular folding could take place in bulk, there is the necessity of
accounting for the change in mechanical properties induced by the
presence of the crystals. One of the principal effects of the crystals is to

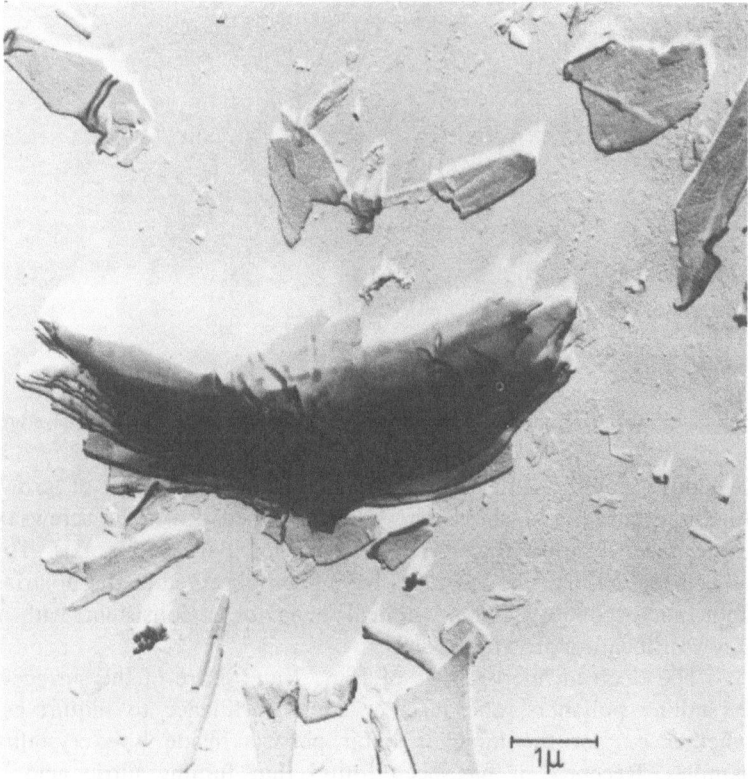

Fig. 9. Electron micrograph of debris from disintegrated polyethylene.
(Courtesy A. Keller and S. Sawada [21].)

Fig. 10. Micrograph of banded spherulites in polyethylene film. Crossed nicols, 200 ×.

increase the toughness and the resistance to shear and mechanical shock. If the crystals were all of the neatly chain-folded lamellae type, it would be difficult to understand how an assembly of such structures could have any strength. It should behave like a deck of cards in shear. Thus, it seems most probable that, although folding is a significant and perhaps dominant process in the formation of the lamellae, there nevertheless must be a considerable number of polymer chains which go from one lamella to the next and form links.

Another aspect of spherulite morphology is illustrated by Fig. 10, which shows the polygonized array formed when a film of polyethylene is completely spherulitized. It should be noted that each spherulite contains, in addition to the black cross, a series of concentric dark bands. It can be shown that these bands are positions of zero birefringence. Furthermore, when viewed on a universal microscope stage between crossed nicols, the bands move and, in some cases, become zigzags along the nicol directions. These observations indicate that along a spherulite radius the optic axes of the crystals twist through an angle that is proportional to the distance from the center of the spherulite [23,24]. Furthermore, over significant areas of the spherulite, this twisting must be in phase (i.e., cooperative) between adjacent radii. While as yet no satisfactory explanation has been given as to why the twist occurs at all, it is interesting to note that electron micrographs of replicas of spherulitized surfaces show a twisting lamellar structure. This is indicated in Fig. 11, where the banded structure is apparent in the surface of a spherulitized film of polyethylene, and

Fig. 11. Electron micrograph of surface of spherulitized polyethylene film. 1000 ×.

Fig. 12. Electron micrograph of the surface of a small portion of the film shown in Fig. 11. 10,000 ×.

also in Fig. 12, which is an enlargement of a portion of Fig. 11. In Fig. 12, the twisting lamellar structure is quite apparent. In a given film, the period of the banding shown in the optical microscope is the same as the period of the twisting shown by the electron microscope. The temperature at which the spherulites are formed alters the periodicity—the lower the temperature, the smaller the period.

Kinetics of Spherulite Growth

Usually the rate at which spherulites grow depends solely on the growth temperature. At a fixed temperature, the radial growth rate is constant. At small supercoolings close to the melting point, decreasing temperature produces a marked increase in growth rate. When some crystallizable polymers (notably those which can be quenched into the glassy state) are allowed to crystallize at very large supercoolings, they exhibit (in addition to the above behavior) a maximum in the growth rate followed by a decreasing growth rate with decreasing temperature. In the region close to the melting point, where the growth rate has a negative temperature coefficient, the magnitude of this coefficient is very large. An example of this shown in Fig. 13, where for this sample of polyethylene oxide the growth rate changed by more than two orders of magnitude with a 7°C change in temperature. The magnitude of this change is strong evidence that the growth of the spherulite is controlled by a nucleation process [25]. If this process is of the two-dimensional coherent type described by equations (1)–(3) [26],

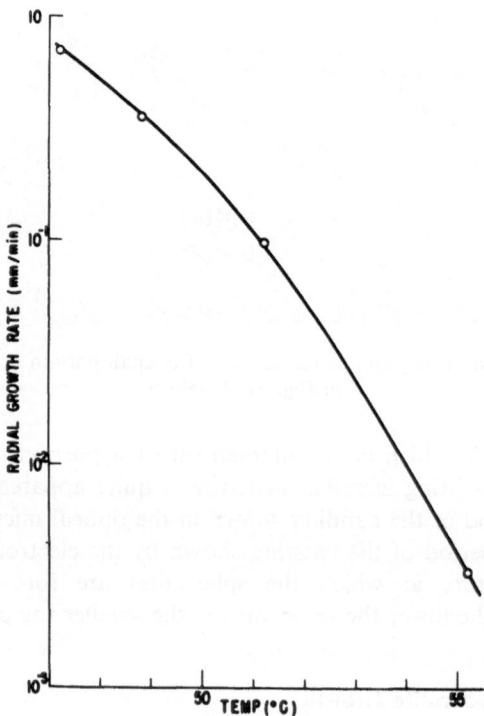

Fig. 13. Influence of temperature on spherulite growth rate in polyethylene oxide (M. W., 12,000).

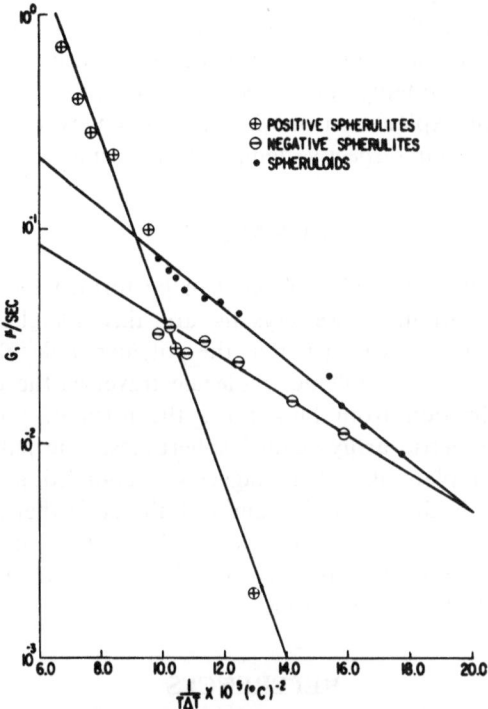

Fig. 14. Plots of log G vs. $1/T\Delta T$ for different kinds of spherulites in nylon 66.

then, as pointed out above, plots of log G versus $1/T\Delta T$ should be straight lines. Figure 14 is one such plot for nylon 66, where the logarithms of the isothermal spherulite growth rates are plotted as functions of $1/T\Delta T$. Data are included for both the positive and negative spherulites mentioned above and, in addition, for some relatively ill-defined predominantly positive agglomerates, called spheruloids [16]. The linearity of the plots for each type of spherulite is satisfactory. Further, it is noted that, if one point is neglected, it is only in a very narrow temperature range—in the neighborhood of the intersection of the lines—that positive and negative spherulites can develop simultaneously. The temperature coefficient of the growth rate of the positive spherulites is very much higher than that of the negative ones. Thus, even if at all temperatures the primary nucleation (birth) rates for positive and negative spherulites were equal, at high temperatures the negative spherulites would predominate, because in this temperature range they grow faster, while at low temperatures the positive spherulites will dominate, for the same reason. It is thus clear that the difference in temperature coefficients of spherulite growth rates yields an *ad hoc* explanation of why the different types of spherulites predominate in their respective temperature regimes.

SUMMARY

Isolable polymer single crystals can be formed by precipitation from dilute solution. These crystals are tiny platelets, the most remarkable feature of which is that the polymer molecules fold back and forth so that each polymer molecule traverses the crystal many times. Crystallization from bulk takes the form of polycrystalline, spherically symmetric arrays called spherulites. The substructure of these arrays in plate-like. This suggests a connection between the crystals formed in dilute solution and in bulk. A further connection is implied by the fact that the growth kinetics and the morphology of both types of crystallization can be fitted into a framework of nucleation-controlled mechanisms.

REFERENCES

1. J. D. Mackenzie, *Modern Aspects of the Vitreous State*, Butterworths (Washington, D. C.), 1960, Chap. 1.
2. P. J. Flory, *Principles of Polymer Chemistry*, Cornell University Press (Ithaca, N. Y.), 1953, Chap. X.
3. L. Mandelkern, *Crystallization of Polymers*, McGraw-Hill (New York), 1964.
4. P. H. Till, *J. Polymer Sci.* **17**: 447 (1957).
5. A. Keller, *Phil. Mag.* **2**: 21 (1957).
6. E. Fischer, *Z. Naturforsch.* **12a**: 753 (1957).
7. P. H. Geil, *Polymer Single Crystals*, John Wiley-Interscience (New York), 1963, Chap. II.
8. A. Keller, *Kolloid-Z.* **197**: 98 (1964).
9. F. P. Price, *J. Polymer Sci.* **42**: 49 (1960).
10. J. I. Lauritzen, Jr., and J. D. Hoffman, *J. Res. Natl. Bur. Std.* **64A**: 73 (1960).
11. M. Volmer, *Z. Physik. Chem.* **102**: 267 (1922).
12. V. F. Holland and P. H. Lindenmeyer, *J. Polymer Sci.* **57**: 589 (1962).
13. A. Keller, *J. Polymer Sci.* **36**: 361 (1959).
14. E. H. Boasson and J. M. Woestenenk, *J. Polymer Sci.* **24**: 57 (1957).
15. F. Khoury, *J. Polymer Sci.* **33**: 389 (1958).
16. J. Mann and L. Roldan-Gonzalez, *J. Polymer Sci.* **60**: 1 (1962).
17. W. Brenschede, *Kolloid-Z.* **114**: 35 (1949).

18. P. H. Geil, *op. cit.*, p. 232 *ff.*
19. F. Danusso and F. Sabbiani, *Rend. Inst. Lombardo Sci. Lettere* **A92**: 435 (1958).
20. R. P. Palmer and A. Cobbold, private communication.
21. A. Keller and S. Sawada, *Makromol. Chem.* **74**: 190 (1964).
22. I. L. Hay and A. Keller, *Nature* **204**: 862 (1964).
23. H. D. Keith and F. J. Padden, Jr., *J. Polymer Sci.* **39**: 101, 123 (1959).
24. F. P. Price, *J. Polymer Sci.* **39**: 139 (1959).
25. P. J. Flory and A. D. McIntyre, *J. Polymer Sci.* **18**: 592 (1955).
26. J. D. Hoffman and J. I. Lauritzen, Jr., *J. Res. Natl. Bur. Std.* **65A**: 297 (1961).

Panel Discussion: Session I

W. G. Courtney (Thiokol): I would like to ask Dr. Hart the following question: Am I right in assuming from what you have been saying that you are not objecting to the cluster concept in nucleation?

E. W. Hart (General Electric Research Laboratory): Oh heavens, no!

W. G. Courtney: Of course not. The objection is to using any term which we characterize as a surface tension, because we cannot define a surface of tension in our cluster. Is that right?

E. W. Hart: Not quite. I rather expect that when the contributions to the actual activation energy for nucleation are put together, they will appear most conveniently as the sum of a volumetric contribution and a contribution proportional to the surface area.

W. G. Courtney: Then, you would say that you could put down a number which would have the form of a surface tension?

E. W. Hart: It will have a form similar to that, but it won't be identifiable with the surface tension of the bulk phase at the same chemical potential.

L. J. Bonis (Ilikon Corporation): Dr. Hart, do you have any comment on Dr. de Bruyn's paper or on any other paper presented in this session? Do you wish to make any additions to your own paper?

E. W. Hart: I think I already made comments on Dr. de Bruyn's paper. There was certainly complete agreement between the viewpoints of Gibbs, de Bruyn, and myself on all essential features. It seems to me that the nucleation problem, in general, does involve a certain amount of confusion. I will make a few general comments on that, although I wish to point out that my main field of interest does not involve nucleation theory as such, nor kinetics in general. Nevertheless, I think it is advisable at all stages to distinguish the parts of the problem. There is, for instance, in each problem of kinetics or nucleation the factor of the particular experimental setup. The system has a particular size, and we know that, when critical nuclei form and grow, the supersaturation of the surrounding material changes, for instance; and, thus, the results obtained really reflect the size of the system, the way in which the heat contact is made, *etc.* However, as far as the fundamental problem is concerned, these factors are irrelevant. I should think one would like to dissociate as much as possible these factors from the main problem, either experimentally or by restriction to the initial nucleation rate, for example (whenever this seems to be reasonably steady-state and not reflecting too much of a transient). I have a little comment to make regarding this particular part of the question. Most of the other discussions on nucleation theory or nucleation were made, of course, with respect to heterogeneous nucleation, and I'd like to emphasize that the results that I claim of zero work of formation do not apply in the case of heterogeneous nucleation.

You might ask why the various forms of representation are based on the current idea of the critical nucleus. My answer would be that, in the first place, they are not exact; they contain many parameters, and even the true answer, whatever that is, will appear in somewhat the same guise. It is still going to be about a system that is spherically symmetric and that contains a critical number of molecules, and thus it can always be written in a form roughly analogous to Gibbs' expression for the work of formation of the equilibrium condensate.

W. G. Courtney: Dr. Russell, please comment upon the effect of transient heat conduction on the actual supersaturation in the cloud-chamber experiment. In my opinion, the transient heat conduction during the time period that the gas is being cooled is sufficient to cause the actual supersaturation in the gas to be several units smaller than the apparent supersaturation.

K. C. Russell (Massachusetts Institute of Technology): Heat conduction from the walls, of course, will reduce the supersaturation ratio below that calculated on the basis of an adiabatic expansion. This decrease results from the following effects: (1) Vapor molecules may transfer energy from the chamber walls to the interior of the system; and (2) a layer of vapor near the walls is heated; then, it expands and recompresses and heats the cold vapor in the center of the chamber. Barnard* found that these effects become important in expansions lasting more than a few hundredths of a second and in chambers of low (< 2 cm) volume-to-surface ratio. He did not find a severalfold decrease in the supersaturation rate by any means.

W. G. Courtney: Please comment upon the appropriateness of relating the volume energy of a cluster to the volume energy of the bulk condensed phase. Note the following: (1) The energy of the molecule in the bulk phase involves contributions from distant atoms due to long-range van der Waals forces; (2) these distant atoms are not present in the cluster; and (3) therefore, the volume energy of the cluster is less than the volume energy of the bulk atoms.

K. C. Russell: The division of the energy of a cluster into volume and surface contributions is, in this case, somewhat misleading. Indeed, in cases where fourth and fifth nearest-neighbor bonds are significant, even the central molecules in clusters may still be affected by the surface. However, molecules the same distance below a planar surface will be similarly affected.

One may define the surface energy as the difference between the energy of the cluster and an equivalent mass of bulk material. Also, it is easy to count the first, second, and third nearest-neighbor bonds broken per unit area of planar surface and cluster surface. Different forms of the attractive potential energy (r^{-6}, r^{-12}, etc.) yield surface energies for curved surfaces either above or below those for planar surfaces.

The bulk surface energy may quite possibly adequately describe the difference

---

* A. J. Barnard, *Proc. Roy. Soc. (London)* **220A**:132 (1953).

between the energy of the cluster and bulk, even though the center of the cluster has properties different from those of the bulk material.

L. J. Bonis: Consider the situation in which there is a solid-state reaction with two possible precipitates. Suppose you have a subcritical particle of A, and then a subcritical particle of B nucleates heterogeneously on it. Will the clusters then be supercritical and grow, or not?

K. C. Russell: This is a rather complex situation, which is unlikely to occur unless phase B has a great tendency to nucleate on phase A, i.e., a very low interfacial energy. If both parts of the double nucleus have radii below their respective critical values, then, in all likelihood, the assembly will decompose. However, it is possible that the addition of the second cluster makes the first particle supercritical, in which case it would become stable and grow.

L. J. Bonis: Do you mean that, in spite of the fact that the double cluster is above the critical size, if either half is subcritical, it could decompose?

K. C. Russell: Yes, the subcritical part could, and this could cause the other part to become subcritical.

J. P. Hirth (Ohio State University): Prof. de Bruyn mentioned the possibility of negative surface energy. I think that there is some indirect evidence for such a negative surface energy in the experiments on the growth of tin whiskers by a mechanism involving growth from the base of the whiskers. Frank's[*] mechanism for this is essentially a mechanism wherein the surface energy of the tin–air interface is negative because tin oxide is stable with respect to tin and oxygen. This negative surface energy provides a driving force for the growth of the whiskers.

P. L. de Bruyn (Massachusetts Institute of Technology): I really have no comment on that, but I think this is a very interesting observation. The question is still whether or not a new phase has been formed. This observation of negative surface tension could indicate that a new phase has been formed.

K. Randolf (Atlantic Research): What is the highest temperature to which your concept may be applied?

J. P. Hirth: For any given system, the limiting temperature for this so-called classical nucleation would be approximately the temperature where the activational free energy for surface diffusion is equal to kT. Above this critical temperature, the high-temperature modifications that I cited would have to be introduced. The critical temperature, of course, will be a material parameter, so that each system must be considered individually.

E. G. Wolfe (Avco Corporation): What does one do about θ and σ in systems where solids, rather than liquid nuclei, would seem to be more meaningful, as in crystal growth experiments? Is θ limited to values characteristic of the crystal lattice, especially since r^* is so small (3–12 Å)?

J. P. Hirth: This question relates to the problem which has come up continually at this symposium of how one thermodynamically describes a small cluster. In the

---

[*] F. C. Frank, *Phil. Mag.* **44**: 854 (1953).

idealized case where the cluster is really some large piece of bulk material, it would be reasonable to discuss the macroscopic contact angle on whatever homogeneous substrate one is dealing with. However, as was evident in the examples I gave, there can be large variations in the effectiveness of the substrate as a nucleation catalyst. In the case of nucleation on ledges, the effective contact angle that one would obtain from a fit to the equations would have no relation to a macroscopic contact angle at all; it would simply be a measure of the constraints of the system with respect to nucleation. In general, for any type of critical-size cluster, we suppose that it has a well-defined free energy of formation. If we subtract from this free energy the bulk free energy change for transformation of the same number of atoms, then the free energy that remains would be ascribed to the surface energy and contact angle. Thus, in a general sense, the contact angle is a phenomenological parameter of the nucleation system.

W. H. Hoback (National Lead Company): Prof. Hirth, you spoke of the influence of a dirty surface on nucleation. When would you consider solid particles dispersed in air to have dirty surfaces? How would this affect nucleation in powder metallurgy, for example?

J. P. Hirth: The only surfaces I would classify as clean would be either ones that were formed by cleavage in vacua better than 10^{-9} mm Hg or metal surfaces that had just been flash-outgassed, again in vacua of better than 10^{-9} mm Hg.

C. H. Li (Grumman Aircraft Corp.): How would you account for the effects of impurities on nucleation? If we have a trace of impurities, what would it affect— the energy of adsorption, the energy of diffusion, or what?

J. P. Hirth: The most important consequence would be the effect on the surface energies, because these appear in the free energy of formation which appears in the exponential in equation (18) of my paper. The secondary effects would be the effect on the desorption energy and on the activation energy for surface diffusion. By and large, the most important effect is the general lowering of the free energy of formation of the critical nucleus by adsorption, which lowers the surface energy. There is some experimental verification for this in nucleation in dilute solutions.*

C. H. Li: It appears that a very slight amount of impurity would affect the behavior greatly, and this would seem to mean that somehow the thermodynamic quantities are discontinuous functions of the impurity concentration at the lower-concentration end. Would you comment on that?

J. P. Hirth: Trace amounts of impurities could be important. For example, we know that trace amounts of oxygen can change the surface energy of silver by a factor of 10^2 or 10^3. However, such effects, in general, would not be discontinuous.

E. W. Hart: There is one way in which without much magic there could be a discontinuous change. Suppose that the impurity was of such a nature that it either formed a saturated adsorbed layer or formed a surface impurity phase at a critical

---

* J. P. Hirth and G. M. Pound, *Condensation and Evaporation*, Pergamon Press, Inc. (Oxford), 1964, pp. 52 and 105.

concentration. This "phase transformation" on the surface could now discontinuously influence the nucleation behavior at the critical concentration.

F. P. Price (General Electric Research Laboratory): I have only one comment and it is quite general. The things that Dr. Hart has been discussing are ideas and concepts which we hope give us a profound understanding of what transpires and which allow us to relate the phenomena in which we are interested to a host of other phenomena. I think that we can consider the parameters that come out of the application of simple nucleation theory as surface tensions. Alternatively, we can simply regard these as parameters which have the dimensions of a tension. This latter view has a certain amount of utility. It is not very satisfying intellectually, but it certainly allows us to correlate and to some extent extrapolate and, thus, is not a totally useless expenditure of energy and effort. Thus, I think there are two things to bear in mind regarding our discussions today. We wish to achieve an intellectually satisfying and deep understanding of these phenomena, but, even if our ability to achieve this understanding fails, we can still use these imperfect theories to practical and useful advantage in the correlation of observations.

B. Caras (Fifth Dimension, Inc.): Have you studied or do you know of others who are studying polymers in your manner, but those polymerized by electrical discharge?

F. P. Price: I know of no one who is doing this sort of work on polymers made by polymerization in an electrical discharge. However, there are a number of other people in the country who are doing work similar to the work that I described here today.

W. H. Hoback: Would you consider the interface that arises when growing a crystal from a solution as being a dirty surface (for instance, your polyethylene crystal from p-xylene)?

F. P. Price: It depends on whether you are talking about hypothetically pure xylene and hypothetically pure polyethylene, or whether you consider the real experiment. I do not think that I have ever seen a clean surface in the sort of experimental work that I have been discussing today. They are always contaminated with at least one, if not many, substances that we normally do not consider to be there. The molecules of the polymers with which we work have a tremendous range of molecular weight. Thus, if you want to consider the low-molecular-weight fractions of these specimens as impurities, I think you are justified in doing so. So, even if you had 100% pure xylene and you used the purest polyethylene you could get, you would still be dealing with a range of materials.

C. H. Li: In connection with impurities, could some of these phenomena observed by Dr. Price be explained on this basis? For example, there's a sharp increase in nucleation rate—several (or is it seven) orders—if the temperature is increased from 47 to 55°. Now, this could be explained on the basis of the formation of a new phase. Therefore, the mass-transfer mechanism has been changed from diffusion to convection or solution. Also, upon annealing some of your spherulites could possibly change size and thereby embrittle the material, analogous to the case of grains growing in metals.

F. P. Price: It seems most improbable that a change in mechanism could be important. Since the molten polymer is very viscous, the mass-transfer mechanism cannot be convection. In reply to the second statement, I would like to say that there have been some reports of spherulite size changes during annealing. I have never seen such changes.

Surface Self-Diffusion at High Temperatures

P. G. Shewmon

Carnegie Institute of Technology
Pittsburgh, Pennsylvania

INTRODUCTION

The first evidence that atoms diffuse much more rapidly over the surface than through the lattice was presented in 1921 by Volmer and Estermann [1], who studied the rate of growth of mercury whiskers from a mercury vapor and found that they could explain their results only if they assumed that atoms hitting the shank of the whisker diffused over the surface to the tip before becoming incorporated in the lattice. In the intervening 35 years, there were many observations that indicated that atoms could diffuse along surfaces much more rapidly than through the lattice. However, these were primarily qualitative observations; the quantitative work was performed after 1955. Some of the work has been done with the field emission technique, for example, work on the spreading of adsorbed gases on tungsten tips [2]. This topic is not discussed here; what is exclusively considered is the diffusion of metal atoms over their own surfaces.

EXPERIMENTAL

The studies of surface self-diffusion that are quantitative started after Mullins published an exact analysis of the surface-tension-driven smoothing processes for particular simple geometries [3]. For example, consider a solid which has a sinusoidal ripple in its surface. A cross section normal to such a surface is shown in Fig. 1. It can be seen that this surface has a higher surface area than the smooth surface, and, therefore, surface tension will tend to smooth it out. Thermodynamic considerations, in the form of the Gibbs–Thomson equation, indicate that the chemical potential, or, equivalently, the vapor pressure, will be higher at the higher parts and lower at the lower parts. Thus, there

111

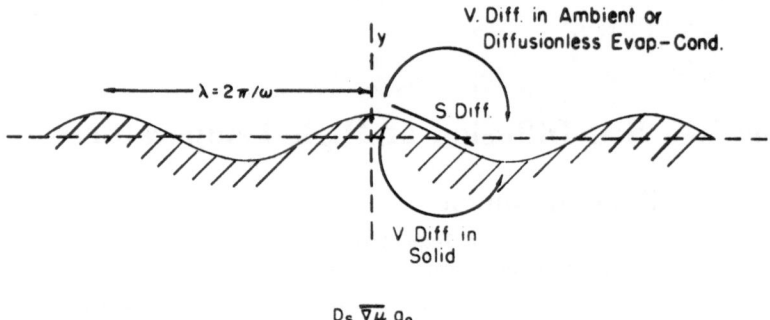

$$J_s/J_v = \frac{D_s \overline{\nabla\mu}\, a_o}{D_v \overline{\nabla\mu}\, \lambda/2} \simeq D_s/D_v\,(\lambda/2a_o)$$

Fig. 1. Section of surface containing sine wave showing the various transport mechanisms that can cause the surface to smooth out.

will be transport of atoms from the high parts to the low parts (from the "hills" into the "valleys"). This action will tend to smooth out the surface and, thus, decrease the total surface area. Figure 1 shows three possible paths by which this diffusion can occur. The first is the diffusion of atoms through the solid, or equivalently the diffusion of vacancies in the opposite direction. The second possibility is diffusion along the surface, that is, in a thin layer which will be about one atomic diameter thick, δ. The atoms on the surface move with a much higher jump frequency than those in the lattice; therefore, this thin surface layer can transport an appreciable quantity of material. Finally, flow can occur by evaporation, diffusion through the gas, and condensation.

The drop in potential between a hill and a valley is the same for diffusion along each of these paths; thus, the mean potential gradient along each of the paths is roughly the same. The rates of flow along the various paths are, consequently, proportional to the diffusion coefficient along that path and the cross-sectional area through which diffusion occurs on the path. The relative quantities flowing over the surface and through the lattice are thus

$$\frac{J_s}{J_l} \simeq \frac{D_s \delta}{D_l \lambda/2} \tag{1}$$

where D_s and D_l are the surface and lattice diffusion coefficients, respectively; λ is the wavelength of the ripple; and δ is the atomic diameter. The vapor pressure of most metals is so low that transport through the vapor can be neglected. Note that with smaller values of λ this ratio increases, and, accordingly, if we have a ratio of D_s/D_l which is appreciably greater than one, there is always some value of λ below which $J_s > J_l$. Experiment shows that, for most metals at high temperatures, the ratio of these diffusion coefficients is of the order of 10^4 or 10^5. As a result, for cases in which the wavelengths λ are less than approximately $10\,\mu$, more material will flow over the surface than through the lattice. For long wavelengths, volume diffusion is dominant.

This demonstrates the basis for all the mass-transport techniques for measuring surface diffusion. It is necessary to work with a small enough distance between changes of sign in curvature in order that surface diffusion be the dominant mode of material transport. Given this general point, there are a variety of experimental techniques that can be used to achieve the rapid change in surface curvature. One method involves putting isolated scratches on the surface. With a field emission tip, the difference in curvature arises between the tip and the less sharply curved shank. Another technique involves the use of a grating machine to produce a series of parallel, equally spaced scratches on the surface.

The technique this author has used most often is that which involves making a grain boundary groove. (Figure 2 shows this geometry.) Where the vertical grain boundary meets the horizontal surface, there is a balance of surface tension—the sort of arrangement seen for interfacial equilibrium where three liquids meet. The same

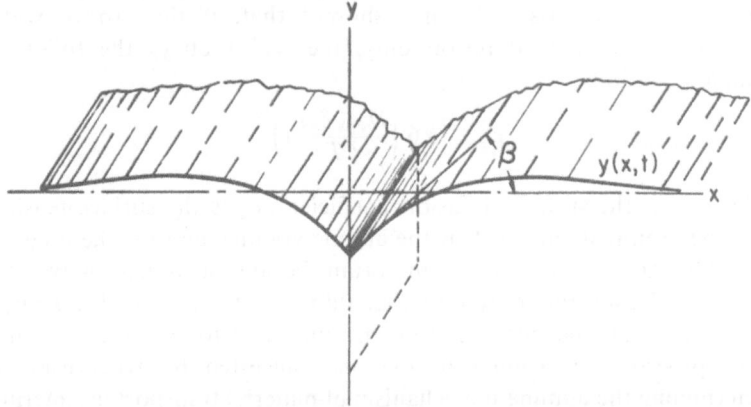

Fig. 2. Section showing ideal groove formed where a grain boundary meets the free surface.

argument applies for the case of solids at high temperatures. The groove angle β is determined by a balance of the grain boundary energy (tension) and the surface tension. It is assumed, and empirically found, that this equilibrium angle is set up very quickly and is maintained throughout the experiment. The sharp groove sets up a curved region on either side of the groove which blends into the smooth surface. As a result, there is transport away from the curved groove root, and this transport plus the fixed angle results in the enlargement of the whole groove. (The shape shown in Fig. 2 is "ideal" in that the surface tension is the same for all exposed surface orientations.) The groove becomes deeper and wider as material is transported out to its smooth surroundings, and consideration of what mechanism moves most of the material gives an equation similar to equation (1), in which the wavelength is replaced by the distance between the maxima on either side of the groove. Thus, given a groove width of approximately 20 μ or less, surface diffusion carries away most of the material; if the groove width is greater than 20 μ, volume diffusion makes a dominant contribution to the process.

With regard to experimental technique, grooves 10–20 μ wide are easily studied with one of the commercially available optical microscopes which also serve as interference microscopes. For grain boundary groove studies, a smooth, strain-free surface is obtained by electropolishing, and then it is grooved by annealing in a reducing atmosphere [4] (usually hydrogen). Some work has been done by annealing in vacuum, but at higher temperatures a gas greatly reduces the evaporation losses. Mullins showed that, if the groove width develops by surface diffusion only, the width obeys the following simple equation:

$$w_s = 4.6 \left(\frac{D_s \sigma_s \, \delta \Omega}{kT} \, t \right)^{1/4} \tag{2}$$

where D_s is the surface diffusion coefficient, σ_s is the surface tension, δ is the atomic diameter, Ω is the atomic volume, and t is the time.

The fact that $w_s \sim t^{1/4}$ is important because a similar analysis for the case of transport only by lattice diffusion gives $w_l \sim t^{1/3}$. Thus, a mechanism can be inferred from the time law for groove widening. This procedure is similar to the one suggested by Kuczynski for determining the dominant mechanism of material transport in sintering. It is desirable to emphasize a difference here, however. In the grooving case, there is an exact solution, and there is no shrinkage or any phenomena present such as occur in sintering. No one has obtained an exact solution to the sintering problem yet.

For the noble metals, every quantity in equation (2) except D_s is known. Thus, if the width is measured as a function of time and

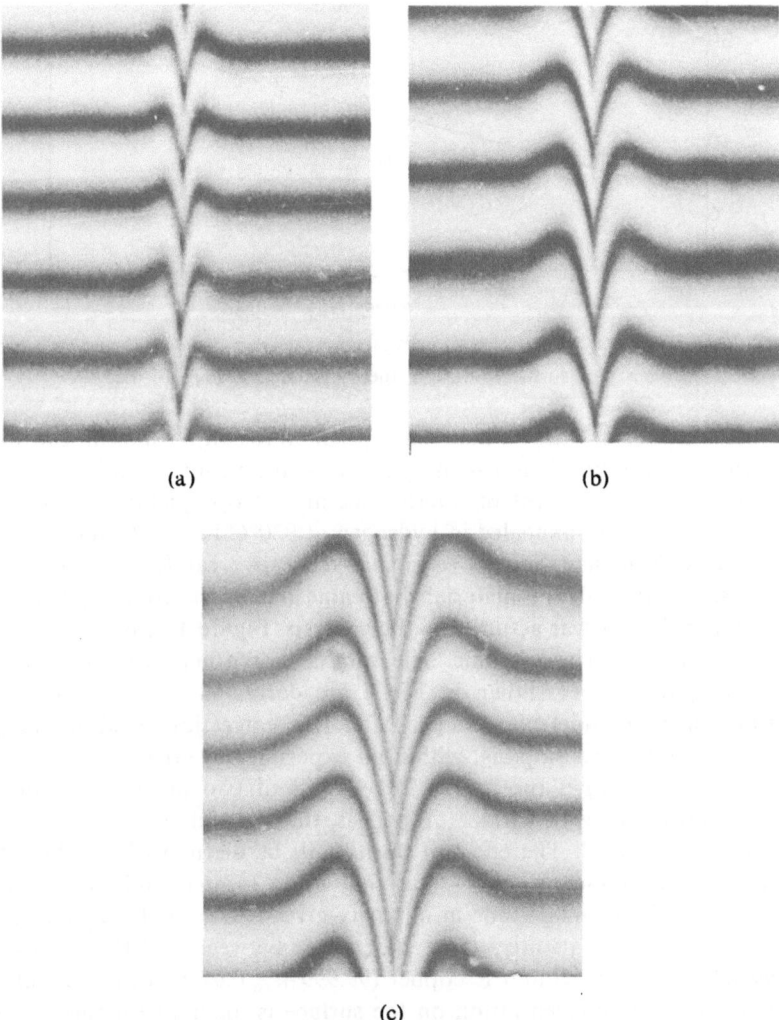

(a)   (b)

(c)

Fig. 3. Growth of grain boundary groove in copper annealed for increasing times in dry hydrogen at 930°C. Photomicrographs taken with an interference microscope. 765 ×. (a) 1 hr. (b) 16 hr. (c) 81 hr.

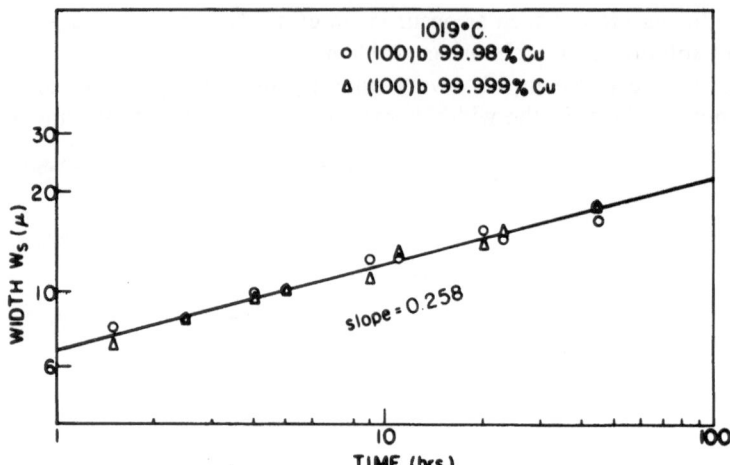

Fig. 4. A plot of log w_s *versus* log t for two purities of copper annealed in dry hydrogen. For surface diffusion, theory predicts a slope of 0.250.

a time law of $t^{1/4}$ is obtained, a surface diffusion coefficient can be calculated. Figures 3 and 4 show some of the results of this type of study. Figure 3 is a set of interference-microscope pictures taken of grooves in copper annealed in hydrogen at 930°C for 1, 16, and 81 hr. It can be seen that the groove increased in size; careful examination also shows that all the material that came out of the root is piled up on either side, as was assumed in the analysis. Figure 4 shows a log–log plot of the width *versus* time. The widths shown here are all $\leqslant 20\,\mu$. Actually, the values shown in Fig. 4 are corrected for a small contribution of volume diffusion [4]. After this correction of about 10% in w, the observed slope is, within an experimental error, 0.250.

Figure 4 shows the results for crystals of two different purities. As in most studies of surface properties, the cleanliness of the surface can be questioned. The effect of purity will be discussed later; but at this time it should be emphasized that using samples differing by at least an order of magnitude in impurity gives results that are indistinguishable. Consequently, any difference between OFHC copper (99.98% Cu) and AS and R copper (99.999+% Cu) is not observable. If the impurity concentration on the surface is different for these two samples, it is not influencing D_s for the copper atoms.

Figure 5 is a plot of D_s *versus* $1/T$ for various copper surfaces. The lines designated (100)a, (110)a, and (111)a represent values of D_s

for surfaces near the (100), (110), and (111) planes, respectively. The main point here is that, when working on any surface, D_s varies by only a factor of 3, or at most, 4; and the activation energy (slope of the lines) is, in essence, the same for all six surfaces. Values of D_s obtained earlier by Gjostein are shown in Fig. 5 as crosses [4]. The agreement in magnitude between the two studies is quite good; however, the slope of the line through Gjostein's points is lower, especially if one considers only his lower-temperature points. [Subsequent figures will show that at lower temperatures the slope (Q_s/R) definitely decreases.]

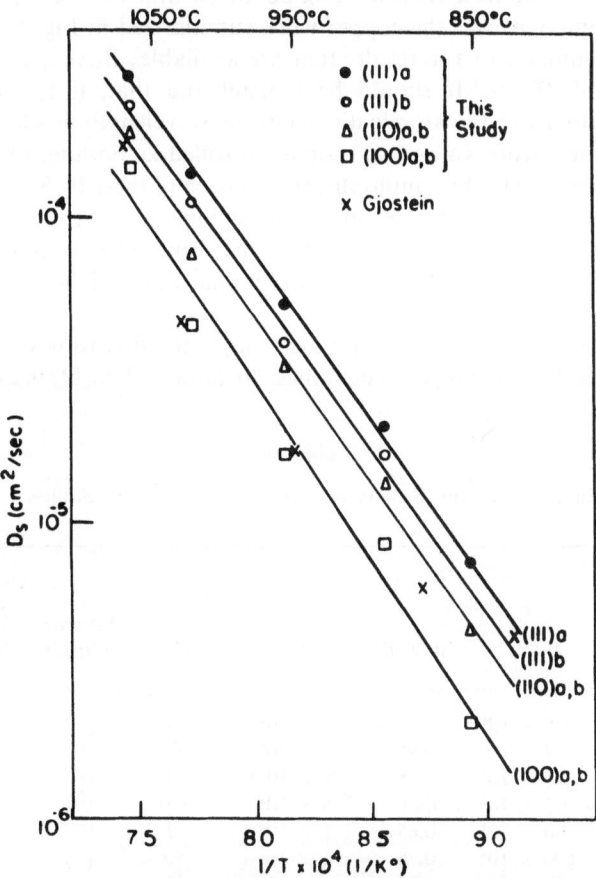

Fig. 5. A plot of log D_s *versus* $1/T$ for various surfaces, showing results obtained from two studies—Gjostein [4] and Choi and Shewmon [5].

If the diffusion coefficient can be given by the following equation:

$$D_s = D_0 \exp\left(-Q_s/RT\right) \qquad (3)$$

the lines in Fig. 5 indicate that $D_0 \simeq 10^4$ cm²/sec, or about 10^5 times the value for lattice diffusion. The valve of Q_s for the lines of Fig. 5 is about 50 kcal, which is about equal to Q for lattice diffusion.

Before discussing these results further, the work done by others should be mentioned. Similar results with high values of Q and D_0 have been found for silver [4,5] and also for gold [6]. Both of these metals have been studied near their melting points (within 15 or 20 % of their melting points), as was the copper work summarized in Fig. 5. Table I shows a summary of the results that are available; these are given as as ratios of D_s/D_l. It should be remembered that, if this ratio is multiplied by twice the atomic diameter, the wavelength at which there is a transition from volume-diffusion-controlled smoothing to surface control is obtained. Thus, multiplication of each of these by 5×10^{-8} cm gives the length that separates these two regions. For copper and gold, this would be about 25 or 20 μ; for silver and nickel, about 15 μ; for α-iron, 50 μ; but, for γ-iron and alumina, much higher transitions are obtained.

Robertson, who worked with alumina, is the first to have successfully applied this technique to ceramics. He annealed Al_2O_3 in air in the

Table I

Summary of the Results of Various Experimental Studies

Metal	D_s/D_l	at T/T_m	D_0	Q_s	References Volume diffusion	Surface diffusion
Cu	4.5×10^4	0.85	2×10^4	49	[24]	[32]
Ag	2.7×10^4	0.85	1×10^6	55	[25]	[33]
Ni	2.8×10^4	0.85	5×10^{-4}	14	[26]	[34]
Au	5.3×10^4	0.85	5×10^4	47	[27]	[35]
α-Fe	10×10^4	0.65	1×10^5	57	[28]	[36]
γ-Fe	800×10^4	0.75	4×10^3	50	[29]	[36]
Pt	1800×10^4	0.77	4×10^{-3}	26	[30]	[37]
Al_2O_3	$10^{11}+$	0.80	6.7×10^2	75	[31]	[38]

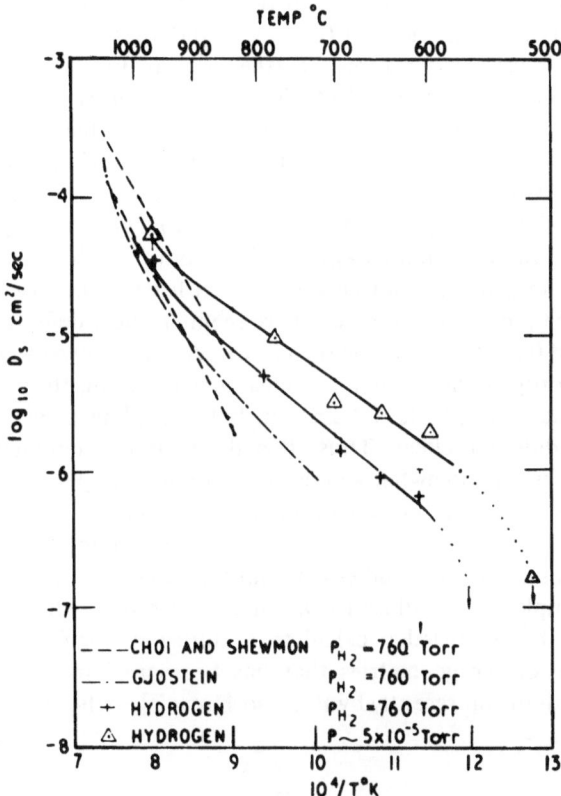

Fig. 6. Values of D_s using various atmospheres over a wider temperature range than shown in Fig. 5. Taken from Bradshaw, Brandon, and Wheeler [6].

temperature range 1700–1100°C and obtained quite satisfactory $t^{1/4}$ kinetics. He did not identify the diffusing species, but this surface process must compete with the diffusion of the slower-moving ion in the lattice (oxygen). The ratio of the surface diffusion coefficient to the oxygen lattice diffusion coefficient is something like 10^{11}; therefore, surface diffusion should be quite important indeed and should completely dominate in the sintering of Al_2O_3, as will be discussed later.

Up to this point, the discussion has been concerned mostly with work at temperatures greater than $0.85\,T_m$, where T_m is the absolute melting temperature. Figure 6 shows some results for copper at lower temperatures [6]. These results, published in 1964 by Bradshaw, Brandon, and Wheeler, agree with their earlier work at high temperatures; but below about 850°C, they found that Q_s is much lower. In fact, at lower temperatures, Q_s (the slope in Fig. 6) can be as low as about one-half that found at higher temperatures. Although there are no satisfactory data for other metals over a wide temperature range, it would fit our meager data to state that this variation of Q_s with T/T_m was valid for most metals. For example, it would allow the nickel data of Blakely and Mykura [36,37] to agree with ours. Their results give an activation energy which is about one-half the lattice Q. These data may or may not be valid, but at least in the case of copper, which is the most studied metal, there seems to be a definite change in Q_s at about 0.85 T_m. It can be seen in Fig. 6 that at lower temperatures and in hydrogen the grooving seems to stop altogether. The exact effect of the hydrogen is not known, but annealing in vacuum has a decidedly different effect. Thus, this discussion is confined to high temperatures—a somewhat simpler case, or so it appears.

To complete the consideration of experimental methods, tracer techniques will now be discussed. As is known, the most common way of performing diffusion studies is to put a radioactive tracer some place on the sample, let it diffuse for a while, find out to where it diffused, and with this knowledge calculate a diffusion coefficient. Figure 7 shows one of the geometries that has been used for this technique in surface diffusion recently by Pye and Drew [7], earlier by Hackerman

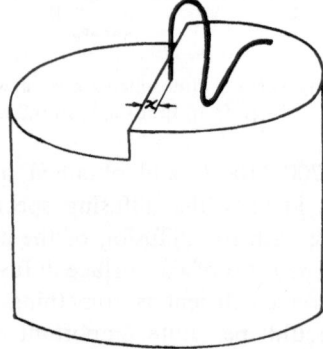

Fig. 7. Specimen geometry for measuring D_s using a radioactive tracer. A wire tip plated with tracer is rested on the surface, the specimen plus the wire are annealed, and then sections are removed. The last slice removed here was a distance x from the wire tip.

and Simpson [8], and by Choi and Shewmon for some copper work [9]. The main reason for considering this is that earlier studies of silver on silver at much lower temperatures yielded very low values of the activation energies (10 kcal) [11]. Silver is a noble metal too, and the results are in marked disagreement with later mass-transport results. There is one basic argument against the procedure used in this tracer work. If you are familiar with the problem of determining a grain boundary diffusion coefficient, you know that a grain boundary can be found only between two pieces of lattice. This is also true for surfaces. Surfaces are always accompanied by an adjoining volume of solid, and this creates experimental problems. If there is a surface source and radio-active material diffuses over the surface, some of the radioactive material will be drained off into the bulk. This is exactly the same thing that happens in grain boundary diffusion where material travels down a grain boundary and continually drains off into the surrounding lattice. What is ultimately counted is the vast majority of the tracer that went into and was trapped in the lattice on either side of the grain boundary. The same thing happens here; the vast majority of the tracer that can be counted after annealing a sample such as that shown in Fig. 7 is trapped below the surface of the metal. This point is emphasized because the investigators who did these earlier tracer experiments completely ignored this fact and stated that everything that diffused out from the source must have stayed on the surface and, thus, only one diffusion coefficient was involved.

This author deems this aforementioned concept invalid, and results obtained using this concept do not correlate at all with those of this author. Since this concept is prevalent in the literature and, thus, the validity of the author's work was questioned (i.e., a low value of Q was assumed correct), this author devised a mathematical solution for the geometry shown in Fig. 7, using the assumption similar to those that Fisher had used for this particular geometry [10,12]. Choi then did some tracer work using a copper substrate and this geometry. Figure 8 shows some of the results that were obtained with copper and gold tracers. The heavier dashed lines in Fig. 8 represent the earlier results for D_s from the grain boundary grooving studies. The grooving and the tracer studies both were done on (100) and (111) surfaces. Since the longest-lived isotope of copper has a relatively short half-life (about 12 hr), a few runs were made using a copper tracer and a fair number of runs were made using radioactive gold, which has a longer half-life (about three days). It can be seen in Fig. 8 that the slope of

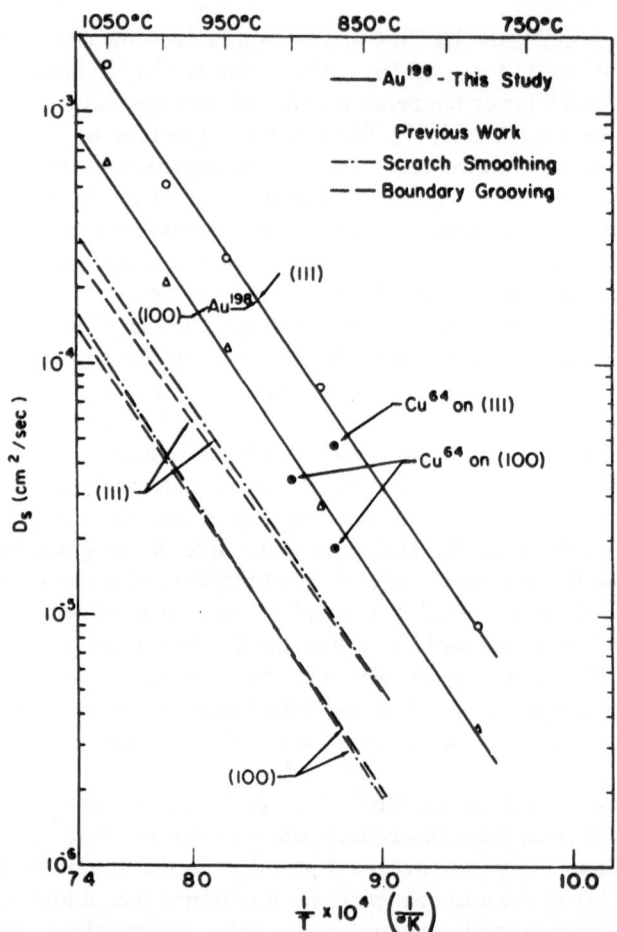

Fig. 8. Surface diffusion coefficients for copper and gold tracers on copper.

D_s for Au[198] on copper is the same as that for Cu[64] on copper obtained in the grooving studies. However, since the actual values of D_s consistently lie about a factor of 4 higher, this author believes this indicates complete agreement with the previously mentioned work. The solution of the differential equation used to calculate this tracer D_s is approximate. Work on grain boundary diffusivities using the exact and approximate solutions usually yields factors of 2 or 3 difference. Thus, the data of Fig. 8 show that the values obtained for D_s for copper with the two techniques are the same within experimental error.

The next consideration is the problem of surface impurities. This problem is discussed widely, but relatively little is known about it. The standard procedure in accounting for divergent results is to cite unclean surfaces. This is a difficult topic to discuss competently, since so little work has been done on the effect of adsorbed impurities on D_s. Using electropolished copper annealed in hydrogen, there is no question but that a fair number of impurities are floating around, even after special rinses and sputtering have been employed. However, field emission tips have really clean surfaces. In view of this problem, the following reasoning was employed: If an impurity is affecting $\dot{D}_s$ and its concentration is changed by an order of magnitude, then the effect on D_s ought to be changed by a large amount. Thus, the purity of the copper is changed by an order of magnitude, and, since it did not measurably affect D_s, it is concluded that impurities adsorbed from inside the metal do not play a dominant role. Accordingly, this author and Gjostein changed the atmospheres with respect to the H_2/H_2O ratio and the H_2/N_2 ratio; the composition of the furnace tubes was changed. In no case did this author find a change in D_s, and Gjostein came to the same conclusion [4]. Thus, our conclusion is that, in the temperature and purity ranges investigated, the impurities present do not affect the measured values of D_s for copper atoms on copper. However, it is admitted that the evidence for this is not as powerful as one might wish.

Although it is not believed possible, assume for a moment that some adsorbed impurities do have an effect on measured values of D_s such that D_s on a *clean* copper surface would be different. The contribution of surface diffusion to sintering will be discussed below. The values of D_s relevant to such mass-transport experiments would be those determined on similar surfaces, while the values of D_s obtained on the *clean* surfaces would be of more value in attempting a theoretical interpretation of the mechanism of diffusion.

Actually, a start has been made on the effect of adsorbed impurities, though little more than this has been accomplished. Rhead has shown that D_s for silver increases with p_{O_2} in the ambient gas [13]. He has recently published preliminary results that indicate that the effect is even more pronounced when annealing is carried out in an atmosphere of varying H_2S/H_2 ratio [14]. In our laboratory, Collins is making a study of the effect of adsorbed sulfur on D_s for copper, a system that has the advantage of a high-temperature adsorption isotherm determined for it at 830°C [15]; thus, there is some information on the chemical nature of the surface. Copper does not facet when sulfur is adsorbed on it; and, therefore, one can still maintain the smoothly curved surfaces of Fig. 2. This is an important point that should be commented on. The techniques discussed for measuring D_s cannot be used if the starting surface tends to break up into low-index planes. To avoid this, one must have something approaching isotropy in the surface tension. This is one reason why the more noble metals were used in this investigation, since they seem to demonstrate this phenomenon. As an extreme case, if one tries to work with something like sodium chloride, one would have problems because any surface will break up into cube faces on annealing; and a shape such as is shown in Fig. 2 cannot be obtained.

Figure 9 shows some of the results obtained for D_s on copper as a function of the H_2S/H_2 ratio of the atmosphere. It also shows the following general effect: Sulfur additions give a rather marked initial increase in D_s which tends to saturate. Some surface orientations do not

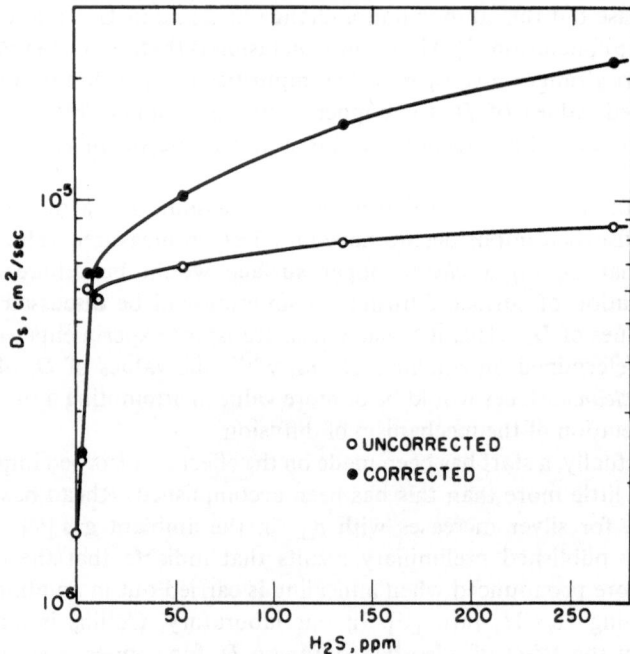

Fig. 9. D_s for copper on a high-index copper surface *versus* H_2S content of the ambient gas. Anneal temperature, 830°C.

come up as precipitously; they come up more parabolically. The two curves of Fig. 9 point out another problem involved with impurities. Since equation (2) includes a surface tension term σ_s, the product $D_s\sigma_s$ is measured, in effect, when the width is measured. To obtain a value for D_s, σ_s must be known. If σ_s for pure copper is used, the curve with the open circles is obtained. If the Bénard and Oudar adsorption isotherm is used and the extent to which the adsorbed sulfur reduces σ_s is estimated, the corrected, upper curve results. This indicates that, rather than saturating, the mobility of copper atoms continues to increase as more sulfur is adsorbed. The maximum H_2S/H_2 ratio shown here is still a factor of 4 below the concentration that gives sulfide on the surface at this temperature.

At this point, a few remarks about field emission work will be made. The procedure here is to form a very sharp tip; clean and anneal it at a temperature that will give blunting; then follow the changes in radius of curvature by shadowing it in the electron microscope. The greatest advantage of this technique is that it permits working with a really clean surface. There are disadvantages, however, one of which is that the work is done primarily with tungsten or, perhaps, molybdenum, which are the metals that have been used primarily to date. It is relatively easy to determine the time required to get a given change in the shadowed radius at different temperatures and, thus, to calculate a value of Q_s. Since it is being shadowed, the exact changes of shape that occur are not known; also there is no exact solution to the diffusion equations involved. Thus, the pre-exponential D_0 is not nearly as accurately established as Q_s. The one metal that has been studied with this method is tungsten [15]. The results given are $D_0 = 4\ cm^2/sec$ and $Q_s = 70$ kcal, which is about half the expected value of Q for lattice diffusion. More recently, Brenner did some work with the field ion microscope in which he studied the changes in the contours of the tungsten tip; he found that the tip really was not circularly symmetric, but that it had undulations around it [17]. This effect results from certain faces developing on the tip. His conclusion was that $D_0 = 10^3\ cm^2/sec$, or about 250 times greater than previously reported. This is the sort of problem that arises. But the surface is clean and the Q_s can be accurately measured. It is hoped there will be more studies, especially on other metals, including some that can be studied by grain boundary grooving also.

With the greater amount of experimental work, there has also been a renewed interest in theoretical models. Although this work is

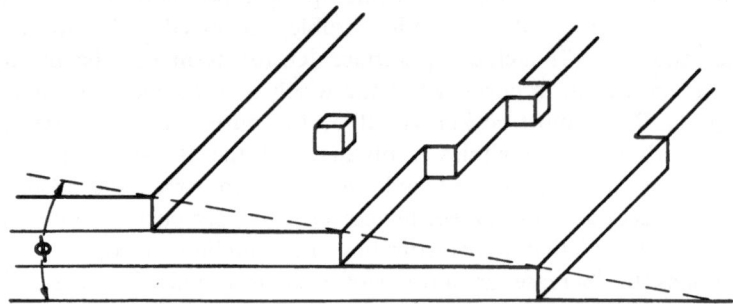

Fig. 10. Schematic of the atomic representation of a surface whose macroscopic orientation deviates from that of a low-index surface by an angle Φ.

tentative, there are two general points worth discussing. The first concerns the activation energies Q_s, which we and others have found for copper, silver, and gold at $T \geqslant 0.85\, T_m$, where T_m is the melting temperature on an absolute scale. When values of $Q_s = Q_l$ were first reported and confirmed for copper, many people were bothered by the high values of Q_s. However, this seems to be common with the noble metals at high temperatures. It is not common, though, for other metals at the lower temperatures that have been used. In discussing the atomic mechanism leading to these large values of Q_s, we have started by assuming the usual ledge–step–kink model shown schematically in Fig. 10. This model certainly is valid for clean surfaces at low temperatures, and perhaps it is also valid at high temperatures. Since no better models are available, this is the one that is usually used.

The next problem concerns the diffusion mechanism, which can be discussed in two ways. It can be said that atoms dissociate from the steps and diffuse over the surface until they get stuck again somewhere else. Birchenall has called this a "rolling-stone" model. It is diffusion *over* the surface. Another model would involve vacancies generated in the surface and, thus, diffusion by a vacancy mechanism *in* the surface. It has been shown that the large values of Q_s, which are independent of surface orientation, can be rationalized by assuming that the atoms are moving over the surface and that the Q_s corresponds to taking an atom from a kink in a step (half-crystal position) out onto the surface and then into an activated position between sites on the smooth low-index surface [18]. In giving the atom this much additional potential energy, one has gone a fair way toward evaporating the atom, and the energy changes involved in evaporating an atom are quite large. Thus, we would argue that it is plausible that taking an atom from a half-crystal position to an activated position diffusing over the surface would require an activation energy that was an appreciable fraction of the heat of vaporization. It is possible to count nearest neighbor bonds and say that it should take $\frac{2}{3}$ of the heat of vaporization. Q_s is indeed fairly close to $\frac{2}{3} \Delta H_{\mathrm{vap}}$. This author would not insist on the $\frac{2}{3}$ at all, however; it is quite possible that, if the diffusing atoms are moving *over* the surface, the Q for this process can be a significant fraction of the heat of vaporization.

Another point to be commented on is the large pre-exponential term D_0. Because of the work of Zener and Wert, it is felt that D_0 for lattice diffusion is fairly well understood. The value of D_0 for surface diffusion on copper is about 10^5 larger than the value found for copper lattice diffusion. This is not really understood; although, after considering various possibilities, the most plausible one is believed to be the following: When an atom on the surface becomes activated, it scoots over the surface until it stops at a new equilibrium position. When working with diffusion inside solids, it has become customary to think of a jump of only one interatomic distance per activation. However, for diffusion *on* the surface (rolling-stone model), there is much less reason to believe that the atom must stop so quickly after it becomes activated. If one postulates that some or all of the activated atoms skate a distance α that is much greater than a_0 each time they are activated, then, since $D_0 \sim a^3$, one can explain the large D_0 [19]. If this hypothesis is accepted, one needs to take $\alpha > 100$ Å to explain the high values of D_0. Since D_0 is independent of surface orientation, this hypothesis requires that an activated atom skate over several steps before it sticks on a new site. This author feels satisfied with this interpretation, but would certainly not be surprised if it ultimately was shown to be wrong.

APPLICATIONS

After this discussion of techniques for determining surface diffusion coefficients and some of the data obtained, some applications of these results will be discussed. Figure 11 shows a sintering model which applies to those initial stages of sintering when spherical surfaces or sections of particles come in contact. The initial stage of sintering involves the formation of joints at these points of contact. An enlarged

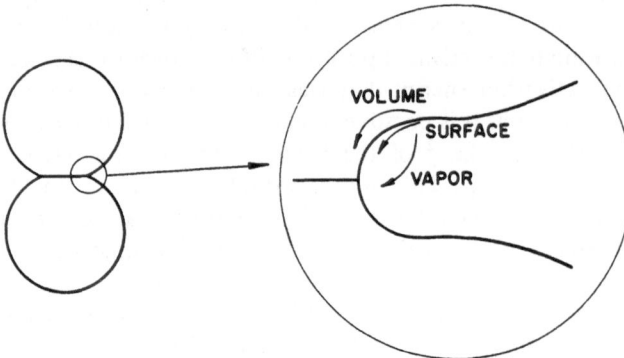

Fig. 11. Cross section of neck formed between two particles in sintering. Enlarged region on right shows three possible transport mechanisms.

section of the neck is shown in the right-hand portion of Fig. 11. Here material is transported from the low-curvature region (particle surface) to the strongly negatively curved region at the neck. The surface area is decreased by transporting matter in this direction, which decreases the free energy associated with the surface. Thus, matter will flow spontaneously in this direction and one need only ask by what mechanism the transport occurs. These mechanisms are the same that were discussed above in relation to Fig. 1.

The mean transport distance in the neck region is of the order of magnitude of the radius of curvature of the neck. This distance invariably is of the order of a micron or less (often much less). Thus, the transport distance is much less than that studied in relation to the grain boundary grooving work. The results in Table I show that, for widths or transport distances that are less than about 20 μ, surface diffusion is dominant. Since this transport distance in sintering is of the order of a micron or less, it must be concluded that surface diffusion clearly is the dominant transport mechanism. On the other hand, when the sintering literature is surveyed, it is generally agreed that volume diffusion is the dominant transport mechanism. However, our disagreement is basic and significant; both arguments cannot be right and this author has a rather definite bias in the matter. Space does not permit discussion of the source of the conclusion favoring volume diffusion, but time laws that were quite approximate were used in the derivation. A more recent computer analysis of this surface-diffusion transport problem has given far more accurate time laws than existed heretofore [20]. In summary, it can be said that all the work that has been done on surface diffusion indicates that, in sintering, surface diffusion has to be the dominant transport mechanism in this early stage of neck growth. The interested reader is referred to the more detailed discussion given by Wilson and Shewmon [21].

One other application concerns the movement of small, inert-gas-filled bubbles inside a piece of metal; this particular application arises in the swelling of fuel elements in nuclear reactors. In fuel elements, each fission event can produce inert-gas atoms. These inert-gas atoms are insoluble, but mobile, and at high temperatures they diffuse around until they combine with vacancies to form bubbles. Since the gas has a negligible equilibrium solubility in the metal, these bubbles cannot coalesce in the way that slightly soluble precipitates do in solids, i.e., by solution of small particles and by diffusion of solute to the bigger particles. However, the bubbles will wander around and combine with one another as a result of collisions, and these collisions give rise to coarsening. The volume of the bubble resulting from a collision is found to be greater than the volume of the colliding bubbles [22]. The pressure difference between the inside and the outside of spherical bubbles is

$$\Delta P = 2\sigma_s/r \tag{4}$$

where σ_s is the surface tension of the bubble and r is the radius of curvature. This equation also applies to bubbles in metals at high temperature. In the metal, the volume (pressure) is adjusted by the addition or removal of vacancies to give the pressure $2\sigma_s/r$. When two bubbles combine, the quantity of inert gas is unchanged; but since

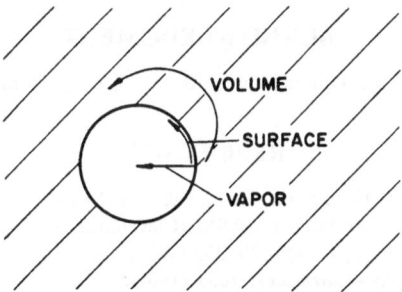

Fig. 12. Diagram showing spherical bubble in solid and the various paths by which matter can be transported around it.

the *r* of the new bubble is bigger, the *equilibrium* gas pressure will be lower. Thus, the volume must increase, by the adsorption of vacancies, until the pressure in the new bubble is reduced and until equation (4) is satisfied by the new radius. As more bubbles combine and the mean diameter increases, the total volume of the gas increases.

It is generally agreed that this bubble growth is the problem involved in fuel-element swelling in nuclear reactors. The rate of wandering of these bubbles is, thus, of importance in fuel-element swelling. In considering the movement of a bubble (Fig. 12), note that the bubble moves by transporting material from one side of the bubble to the other. This can occur by the same three mechanisms discussed previously—surface, volume, and vapor diffusion. An analysis of the relative flow by the various mechanisms leads to an equation similar to equation (1). The effective diffusion distance is the bubble radius, which is of the order of 0.1–0.01 μ. Thus, surface diffusion clearly should be the dominant transport mechanism. Indeed, a more sophisticated analysis of some of the electron microscope observations has proven that this is the case [23].

CONCLUSION

The techniques that have been used for determining D_s have been discussed. There is no question but that a surface diffusion coefficient can be measured and that D_s for the surface atoms is much greater than D_l for atoms in the lattice. Some of the results have been discussed, primarily those obtained at high temperatures; the models which may be applicable in these cases have also been considered. Finally, some of the practical applications of these results have been described.

ACKNOWLEDGMENT

The support of the United States Atomic Energy Commission is gratefully acknowledged.

REFERENCES

1. M. Volmer and I. Estermann, *Z. Physik* 7: 1–13 (1921).
2. G. Ehrlich, in: *Metal Surfaces*, ASM (Cleveland), 1963, p. 221.
3. W. W. Mullins, *J. Appl. Phys.* 28: 335 (1957).
4. N. Gjostein, *Trans. AIME* 221: 1039 (1961).
5. J. Choi and P. G. Shewmon, *Trans. AIME* 224: 589 (1962).
6. F. J. Bradshaw, R. H. Brandon, and C. Wheeler, *Acta Met.* 12: 1057 (1964).
7. J. J. Pye and J. B. Drew, *Trans. AIME* 230: 1500 (1964).

8. N. Hackerman and N. H. Simpson, *Trans. Faraday Soc.* **42**: 376 (1956).
9. J. Choi and P. G. Shewmon, *Trans. AIME* **230**: 123 (1964).
10. P. G. Shewmon, *J. Appl. Phys.* **34**: 755 (1963).
11. R. A. Nickerson and E. R. Parker, *Trans. ASM* **42**: 376 (1950).
12. J. C. Fisher, *J. Appl. Phys.* **22**: 74 (1951).
13. G. E. Rhead, *Acta Met.* **13**: 223 (1965).
14. G. Rhead and J. Perdersan, *Compt. Rend.* **260**: 1929 (1965).
15. J. Oudar, *Compt. Rend.* **249**: 91 (1959).
16. J. P. Barbour, F. M. Charbonnier, *et al.*, *Phys. Rev.* **117**: 1452 (1960).
17. S. Brenner, in: *Metal Surfaces*, ASM (Cleveland), 1963, p. 305.
18. P. G. Shewmon and J. Choi, *Trans. AIME* **224**: 589 (1962).
19. P. G. Shewmon and J. Choi, *Trans. AIME* **227**: 515 (1963).
20. F. A. Nichols and W. W. Mullins, *J. Appl. Phys.* **36**: 1826 (1965).
21. T. L. Wilson and P. G. Shewmon, *Trans. AIME* **236**: 48 (1966).
22. P. S. Barnes and D. J. Mazey, *Proc. Roy. Soc.* (*London*) **A275**: 47 (1963).
23. P. G. Shewmon, *Trans. AIME* **230**: 1134 (1964).
24. A. Kuper, H. Letow, L. Slifkin, and C. Tomizuka, *Phys. Rev.* **98**: 1870 (1955).
25. C. Tomizuka and E. Sonder, *Phys. Rev.* **103**: 1182 (1956).
26. R. Hoffman, F. Pickus, and R. Ward, *Trans. AIME* **206**: 483 (1956).
27. S. Makin, A. Rowe, and A. LeClaire, *Proc. Phys. Soc.* **70B**: 545 (1957).
28. Y. Adda and A. Kirianenko, *Compt. Rend.* **247**: 744 (1958).
29. V. Lyashenko, *Fiz. Metal. i Metalloved.* **7**: 362 (1959).
30. G. V. Kidson and R. Ross, *Radio Isotopes in Scientific Research*, Vol. I, Pergamon Press (Oxford), 1958, p. 185.
31. Y. Oishi and W. D. Kingery, *J. Chem. Phys.* **33**: 480 (1960).
32. J. Y. Choi and P. G. Shewmon, *Trans. AIME* **224**: 589 (1962).
33. G. E. Rhead, *Acta Met.* **11**: 1035 (1963).
34. J. M. Blakely and H. Mykura, *Acta Met.* **11**: 399 (1963).
35. N. A. Gjostein, Ford Scientific Lab, private communication.
36. J. M. Blakely and H. Mykura, *Acta Met.* **11**: 399 (1963).
37. J. Blakely and H. Mykura, *Acta Met.* **10**: 565 (1962).
38. W. Robertson and R. Chang, "The Kinetics of Grain-Boundary Groove Growth on Alumina Surfaces," *Materials Science Research*, Vol. 3, Plenum Press (New York), 1966, pp. 49–60.

Self-Diffusion on Nearly Pure Metallic Surfaces

C. E. Birchenall and J. M. Williams

Department of Chemical Engineering
University of Delaware
Newark, Delaware

INTRODUCTION

This paper presents a discussion of the same topic covered by Shewmon,* rather argumentatively, because we often agree on the premises and differ on the conclusions. Fortunately, he has outlined very nicely the background of many of the topics that are discussed here. We shall repeat some of the results of others that he cited and add a few additional ones because it is our purpose to emphasize those things that are in disagreement. It is not our intention to present this topic as a closed case as he did, because we feel that the case is far from closed. Surface self-diffusion mechanisms are discussed first—a topic which Shewmon also considered.

SURFACE SELF-DIFFUSION MECHANISMS

Different crystallographic planes of a metal differ in their surface energies, in the packing densities of the atoms, and in the equilibrium densities of facets, ledges, and jogs. It is to be expected that the equilibrium concentrations of surface point defects, that is, vacancies, rolling stones (the surface analog of the interstitial), and adsorbed impurity atoms differ as well. Because the configurations of both the equilibrium and saddle-point positions during a unit diffusion jump differ from plane to plane, the enthalpies of motion must differ. Although a single enthalpy of formation must exist for each kind of crystallographic plane, for those planes lacking three- or fourfold rotational symmetry, the enthalpy of motion may vary with direction

* See P. G. Shewmon, "Surface Self-Diffusion at High Temperatures," this volume, pp. 111–131.

within the plane. Observation of diffusional anisotropy within a single plane is evidence of a difference in enthalpy of motion, if temperature-dependent, or a difference in frequency factor (entropy of motion, jump distance, and vibration frequency) if nearly independent of temperature.

It is generally conceded that surface self-diffusion occurs by the motion of individual point defects (although it is probably unwise at this stage to rule out a divacancy mechanism at lower temperatures).

Gjostein [1] proposed that isolated atoms form on otherwise completed planes by dissociation from ledges and migrate by an activated mechanism, as Shewmon discussed. Shewmon and co-workers [2] have extended this model, and their calculations and measurements agree that an activation energy close to that for volume self-diffusion in the same crystal is required. Although for the original model it was assumed that the jumps were short, the measured frequency factors were very high; and it seems to be necessary that the mean free path be relatively large if the mechanism is to account for any of the self-diffusion results. (Hirth [3] pointed out that, for the conditions of long mean free path, it is necessary for the activation energy to be of the order of kT or less.)

Because of this requirement, diffusion by the rolling-stone mechanism on a (110) surface of an fcc metallic crystal should be highly anisoptropic. There should also be adsorbable impurities capable of combining with and immobilizing these relatively reactive atoms. This field has not yet been very fully explored.

Birchenall [4] proposed that, because of the ease of relaxation about a surface vacancy, its enthalpy of formation by releasing an atom to add to a ledge should be much lower than the enthalpy of formation of a rolling stone, since the gross change in coordination is the same. Consequently, the vacancy concentration should be much higher in the surface than that of rolling stones.* The problem is analogous to the problem of the mechanism of self-diffusion in bulk metal where the interstitials, once formed, have a high mobility; however, there are not many of them relative to the vacancy concentration. It should

---

* It should be recalled that Müller [5] found only about as many vacancies on platinum tips in a field-ion microscope as the volume equilibrium concentration predicts. Although these vacancies were observed on successive surfaces exposed by field evaporation, there is little reason to believe that the concentration so determined should be characteristic of the equilibrium surface. In addition, the statistics were relatively few.

be noted also that measurements indicate that the bulk concentration of vacancies in copper, silver, and gold at their melting points is of the order of 1 in 1000 to 1 in 10,000 sites [6]. It is difficult to believe that the concentration of surface vacancies is not appreciably higher. In the extreme case, when half the surface sites are vacant, the other half are rolling stones, and the two models become indistinguishable. In other words, a continuous transition could take place at very high defect densities. Because the surface location relaxes the steric interference found in the motion of an internal vacancy, the enthalpy of motion should be low for a surface vacancy mechanism. Furthermore, there seems to be little reason to expect large anisotropy on surfaces like the (110) of an fcc metallic crystal. The only kind of adsorbable impurity likely to immobilize the vacancies completely is an atom small enough to fall into and annihilate the vacancy.

METHODS OF MEASUREMENT OF SURFACE SELF-DIFFUSION COEFFICIENTS

There are two basic types of experimental procedures employed in attempts to measure surface self-diffusion coefficients—shape change and tracer spreading.

Shape Change

The shape-change methods involve a surface-energy driving force. A variety of configurations and shape-determining devices has been tried. Kuczynski [7] studied the changes in the junction of a small sphere resting on a plate. He and his followers later studied the behavior of strings of spheres in contact, wires wound on a spool, and wires twisted together, by measuring the junctions by optical microscopy. These measurements were always carried out close to the melting point of the metal and were usually interpreted in terms of a volume-diffusion vacancy mechanism as the rate-controlling process.

Some of the numerical agreement reported between the directly measured (tracer) volume diffusion coefficients and the sintering (neck-growth) experiments is surprisingly good. In passing, note that the important wavelength governing mass transport in this measurement is not the particle size, but rather the radius of curvature in the growing neck.

Udin *et al.* [8], in mechanically measuring the surface tension of wires, noticed that grooves developed at the grain boundaries. After Mullins [9] analyzed the kinetics of the grooving process and later the related process of smoothing of scratches, Gjostein, Shewmon, Blakely, Mykura, and others utilized the method. They determined the surface contours by optical interferometric microscopy. Most of the measurements were performed very close to the melting point and were interpreted almost entirely in terms of a surface-diffusion mechanism as the rate-determining step, even when the materials and the temperatures were the same as those used in the neck-growth experiments. Some recent measurements [10] have been done at intermediate temperatures as well. In reviewing the data Shewmon reviewed, we have emphasized the disagreement, whereas he emphasized the agreement of one set of these measurements with other types of measurements.

Shape changes of field-emission points [11] can be followed by changes in the electron-emission pattern in the field-emission or field-ion microscope. These measurements can be done only on metals able to withstand the stresses in the field-emission system. Relative to the melting points of these metals, the temperature range in which surface-mobility measurements are made is usually lower than that for the preceding shape-change methods. Because the tip usually approaches a complete hemisphere, nearly all orientations of the crystal are present somewhere. Some faceting always takes place. Presumably, the atoms move most rapidly across the flat faces in order to preserve the facet. In comparison with this needle-blunting method, a similar range of orientations is present in the neck-growth procedure (it also has one grain boundary in the neck). However, in the boundary-grooving or scratch-smoothing procedure, all the planes exposed lie within approximately a 10° rotation about a single crystallographic axis that lies parallel to the scratch or to the boundary intersection with the surface. Facets have been reported [10] on the sides of these grooves, again indicating an orientation dependence of the surface diffusivity.

In all of these shape-change methods, the atoms on the surface and those below the surface are indistinguishable. Exchanges occurring between surface and volume are unobservable. Such exchanges do not affect these measurements. Thus, we can learn little about surface–volume exchanges from shape-change methods.

Evaporation from the surface probably does not interfere seriously with the grooving and scratch-smoothing measurements, unless the rate of evaporation differs substantially with location in the critical area, or unless the vapor recondenses in a preferential way to contribute to shape change. The particles used in neck-growth and field-emission experiments may be sensitive to evaporation losses when the relative temperature is high, simply because of the relatively small sizes of the particles. The heats of sublimation for metals are greater than the activation energies for volume self-diffusion and, therefore, are greater than the highest values attributed by anyone to surface diffusion. Consequently, the rate of vaporization of the surface atoms is greater, relative to surface mobility, the higher the temperature for a given substance. Table I gives approximately the number of monolayers of metal evaporated in one hour at 0.6, 0.8, and 1.0 T_m (where T_m is the absolute melting temperature) of the metals on which surface self-diffusion measurements are reported.

Near the melting point, in an hour, which is a reasonable time for a measurement of this kind, something of the order of 10^3–10^7 monolayers might evaporate if the system is not maintained in equilibrium with respect to metal vapor. Therefore, in shape-change measurements made at high temperatures, it is essential that the system be truly in equilibrium with respect to the vapor phase and that there be no parts

Table I

Maximum Number of Monolayers Evaporated in 1 hr from a Clean
Metal Surface into Vacuum*

Metal	Monolayers per hour			$T_m(^\circ\text{K})$
	$0.6\ T_m$	$0.8\ T_m$	T_m	
Ag	5.1×10^{-3}	4.2×10^2	4.1×10^5	1233
Au	7.5×10^{-7}	3.8×10^{-1}	9.9×10^2	1336
Cu	1.2×10^{-4}	3.5×10	7.0×10^4	1356
Ni	4.0×10^{-3}	6.3×10^2	8.5×10^5	1728
Fe	2.0×10^{-1}	1.5×10^4	7.5×10^6	1808
Pt	1.2×10^{-5}	9.0	1.9×10^4	2047
W	1.4×10^{-1}	6.3×10^3	2.8×10^6	3743

* Based on an assumed evaporation coefficient of unity. Data from S. Dushman, *Scientific Foundations of Vacuum Technique*, John Wiley & Sons (New York), 1949, p. 745.

of the system accessible to the vapor cooler than the working surface itself. This is clearly a potentially important source of error. On the other hand, tracer-spreading measurements are usually made at such low temperatures that vapor transport does not appear to be a likely explanation of any discrepancies.

Tracer Spreading

A method distinctly different from the shape-change procedure involves the spreading of a radioactive tracer from a (hopefully) well-defined source onto a surface of previously determined orientation. The driving force for spreading is the entropy of mixing rather than the difference in chemical potential resulting from the curvature. On a plane surface, there should be no contribution from surface-energy variation. As in the shape-change studies, a variety of configurations and mass-transfer-detection schemes has been employed in the tracer experiments.

Nickerson and Parker [12] studied the self-diffusion of silver by following the spreading of a radioactive isotope along a bundle of fine, polycrystalline wires. Their tracer detection system was a lead-collimated geiger tube arranged so as to permit the diffusion specimen to be scanned without breaking the vacuum (10^{-6} torr) on their apparatus.

This work was followed with another investigation of silver by Winegard and Chalmers [13], who utilized autoradiography to ascertain the redistribution of the tracer material. Although no activation energy could be obtained from their results to compare with that of Nickerson and Parker, the variation of the diffusion rate with orientation and direction was positively established.

Another detection procedure, scintillation counting, was employed by Hackerman and Simpson [14] in their study of copper, where a needle was used as the tracer source. No activation energy was obtained, but the existence of anisotropy was reported again on a symmetrical surface on which it should not have appeared if the surface had been ideally oriented.

The detection technique of sectioning and counting as employed by Geguzin et al. [15,16] has led to the majority of the reported results of tracer diffusivities. Unfortunately, it requires destruction of the diffusion specimen, whereas with the other detection methods the tracer distribution can be obtained in situ.

Thus, perhaps the most generally useful detection technique is autoradiography, for it affords the greatest resolution of the tracer distribution and certainly elucidates anisotropic behavior most clearly. The tracer distribution can be determined before and after the diffusion anneal. Additionally, penetration into the bulk of the specimen can be simply ascertained by slicing the specimen at a very small angle to the original surface and making an autoradiograph of this new surface.

Naturally, a correlation correction to the tracer diffusivity is required for surface diffusion, as it is for volume diffusion, so as to convert it to the self-diffusion coefficient if a vacancy mechanism is responsible. In the present state of the art, this is too small a correction to cause much concern.

The tracer atom is distinguishable from all other atoms, and it is possible, in principle, to observe the process of exchange between tracer atoms from the surface and stable atoms from the interior. In fact, this procedure has been used to study the volume-diffusion rate for a long time (the method of decrease in surface activity). It has often been assumed that once an atom moves any distance beneath the surface, it loses its lateral mobility completely. However, it has not been demonstrated that the surface is always as shallow as one or two layers, and the Russian investigators especially have begun to question the validity of this assumption. The thickness parameter δ plays an important role in the mathematical solution to the tracer-spreading problem when losses to the volume are involved. This problem is totally analogous to the grain boundary diffusion problem, apart from some differences in the initial geometry, as it was defined by Fisher [17]. Geguzin, Kovalev, and Ratner [16] have even synthesized a surface analog to Fisher's grain boundary diffusion model that reproduces the geometry exactly. They have worked out a more detailed solution that differs from that given by Whipple [18] and have experimental data for surface self-diffusion on gold and iron that seem to support their solution.

By means of a Fourier–Laplace transformation, Whipple has obtained the exact solution for the grain boundary penetration problem. Geguzin and co-workers [16] obtain another solution via an unusual contour integration and subsequent approximation. This solution is of exceptional interest because it purports to permit both the surface diffusivity D_s and the thickness of the high-mobility layer δ to be determined by appropriate manipulation of the experimental concentration distribution in a single experiment.

All other mathematical treatments allow only the product $D_s\delta$ to be determined. Either D_s or δ must thus be obtained independent of the diffusion study. Through judicious treatment of the two solutions, they can be shown to be essentially identical, with the exception of the term which permits Geguzin to evaluate the two parameters separately. Therefore, future effort should be directed at establishing the validity of Geguzin's solution because of its apparent usefulness. Pavlov and Panteleev [19] have shown that the surface self-diffusion of iron was measured by Geguzin et al. under conditions that lie outside the valid range for their mathematical solution. Their reported values must be considered doubtful.

Recognizing the increased importance of accurately accounting for tracer losses into the bulk during a surface-diffusion study, several investigators have attacked the mathematical problem in various manners. The significance of such surface losses via lattice diffusion or diffusion down dislocations or grain boundaries has been pointed out by Blakely [20]. Shewmon [21] has obtained an approximate solution to the the trapping problem based on the assumptions of Fisher's grain boundary analysis. His results are particularly appropriate for the experimental technique of sectioning and counting, but the initial assumption of a steady-state surface concentration casts doubt on the detailed usefulness of the subsequent result. Another approach to the problem has been taken by Drew and Amar [22], who solve the one-dimensional diffusion problem with no trapping and then use it as the boundary condition to determine a correction for losses to be applied to the "no trapping" surface concentration. Their resulting equation is applicable to experimental techniques, such as autoradiography, which measure the surface concentration. Although it indicates the influence of volume effects, it must be regarded as only a first approximation. A third solution to the trapping problem has been developed by Williams [23] using the physical model of both Shewmon and Geguzin. Without invoking any assumptions about the transient behavior of the tracer on the surface, an exact expression for the surface concentration was derived. Unfortunately, the result is in the form of an integral, and, indeed, it can be shown to be equivalent to Whipple's; however, the significant trapping parameters—the penetration depth $(D_v t)^{1/2}$ and surface thickness δ—are the same as those in Drew and Amar's treatment. An application to the experimental silver results of Drew and Pye [24] is consistent with the observed temperature dependence of the diffusivity.

EXPERIMENTAL SURFACE SELF-DIFFUSION RESULTS

Table II is a summary of nearly all of the surface self-duffusion data reported for metals. The methods and temperature ranges are also given.

Birchenall [4] pointed out that these surface self-diffusion measurements on ostensibly clean surfaces yield apparent activation energies that correlate in a rough way with the range of temperature relative to the melting temperature in which the measurements were carried out. Because grooving and scratch-smoothing are used at the highest temperatures, field-emission at intermediate temperatures, and tracer spreading mainly at the lowest temperatures, the apparent activation energies also correlate roughly with the method of measurement.

However, there are certain agreements and disagreements to observe. First, in the case of silver, there are activation energies ranging from 8 to ~56 kcal/g-atom, which indicates a slight discrepancy. The low values are all associated with tracer methods, whereas the high value is associated with a grooving experiment. The temperature ranges are of the order of one-half the melting temperature for the tracer experiments and about 0.8 of the melting point in the grooving measurement. In the case of copper, the spread of activation energies is similar, but not quite as large, ranging from an estimated value of the order of 13 to the more firm values from about 17 to 55. And again there is correlation with the temperature range of measurement. In this case, however, agreement in magnitude between the tracer and grooving experiments near the melting point has been reported by Shewmon and co-workers. The nickel results are especially interesting. Blakely and Mykura [10], using smoothing techniques, report two rather different values for different surfaces. They attribute the higher activation energy on the (100) surface to impurity effects; therefore, the activation energy to which they give the greater weight is about 14 kcal. Pye and Drew [28], making tracer measurements on these same surfaces, find diffusivities that agree very well numerically; that is, there is a factor of only 2 or 3 difference. Also, the activation energies agree very well. Thus, here is a case in which there is overlap in the temperature range; two entirely different experimental methods have been employed; the numerical values agree very well indeed. Even the (100) value of the diffusion coefficient for smoothing intersects the (100) and (110) values from tracer measurements, although the slopes are different.

Table II

Summary of Surface Self-Diffusion Measurements

Element	Plane	D_0 (cm²/sec)	Q_s (kcal/g-atom)	Temperature range (°C)	Relative temperature range	Method	Reference
Ag	(321)	0.2	10.3	250–400	0.42–0.55	Tracer	[12]
		1.5×10^{-5}	8.1	300–500	0.46–0.62	Tracer	[24]
		10^6	55.5	700–945	0.79–0.99	Grooving	[35]
		0.3	11.8	270–695	0.44–0.78	Tracer	[15]
Au		7.6	15.7	272–612	0.41–0.66	Tracer	[36]
			10–20	500–950	0.58–0.91	Tracer	[26]
			16–40			Grooving and smoothing	[26,27]
			$\sim 23 \pm 5$	557–630	0.62–0.68	Field-emission	[37]
Cu	(110)		$\sim 13 \pm 2$	700–1050	0.72–0.98	Shape change	[38]
	(100)	6.5×10^2	40.8	720–1070	0.73–0.99	Grooving	[1]
		2×10^4	49	847–1069	0.83–0.99	Grooving	[2]
	(100)	0.07	19	700–950	0.72–0.90	Smoothing	[39]
	(111)	10.1×10^4	54.5	850–1060	0.83–0.98	Smoothing	[2]
		4.4×10^4	50.3	850–1060	0.83–0.98	Smoothing	[2]
	(110) to (112)	0.03	17	400–800	0.50–0.79	Smoothing and grooving	[25]

Table II (*continued*)

Element	Plane	D_0 (cm²/sec)	Q_s (kcal/g-atom)	Temperature range (°C)	Relative temperature range	Method	Reference
Ni	(111)	5×10^{-4}	14.3	800–1200	0.62–0.85	Smoothing	[10]
	(100)	0.7	39.2	800–1200	0.62–0.85	Smoothing	[10]
	(111)	6.5×10^{-5}	13.8	400–1000	0.39–0.74	Tracer	[28]
	(111)	1.8×10^{-5}	13.8	400–1000	0.39–0.74	Tracer	[28]
γ-Fe		4×10^3	51	910–1100	0.65–0.76	Grooving	[40]
α-Fe		10^5	57.8	750–910	0.57–0.65	Smoothing	[40]
α-Fe		1.5×10^2	31	550–860	0.46–0.63	Tracer	[16]
Pt		4×10^{-3}	25.8	890–1310	0.57–0.78	Smoothing	[41]
W		4	72	1530–2430	0.40–0.64	Field-emission	[11]
W		4	64.3	1430–1830	0.38–0.48	Field-emission	[42]
Re			34.5	930–1530	0.35–0.52	Field-emission	[43]

This agreement between different methods in this lower temperature range favors a relatively low activation energy.

In order for this field to develop fruitfully, it is important to determine the cause of the apparent discrepancies in the surface self-diffusion data. Several possibilities emerge, but none seems to be clearly established. Some possibilities are listed below for any good that they might do for those who are inclined to make measurements in the future.

It is possible that the different methods measure somewhat different processes. Bradshaw et al. [25] suggest that, in the lower part of the temperature range (the copper results discussed by Shewmon elsewhere in this volume) in which they examined grooving or smoothing on copper, a volume-diffusion contribution played a role (in hydrogen atmosphere).

Actually, their low activation energies for surface self-diffusion in the intermediate range would be more compatible with a volume contribution at higher temperatures. Otherwise, there is difficulty in reconciling the neck-growth measurements with the grooving experiments, as Shewmon also pointed out. Making a precise determination of the exact proportions of $t^{1/3}$ and $t^{1/4}$ contributions to grooving or smoothing must require phenomenal experimental precision.

For surface self-diffusion measurements on gold, Geguzin et al. [26] show a continuously increasing activation energy with temperature for both grooving and smoothing measurements (they include Blakely's [27] results with theirs). Figure 1 shows the results of Geguzin et al. [26]. Their tracer results (the upper curve) lie nearly parallel to, but two orders of magnitude higher than, the results for grooving and smoothing. However, the slopes seem to be approximately equal. The slope ranges from about 10 kcal at the low temperatures to about 40 kcal at the highest temperatures. (On the other hand, Blakely and Mykura's [10] nickel smoothing results agree in magnitude very well with Pye and Drew's [28] tracer measurements.)

Bradshaw et al. [25] report essential deviations from the Arrhenius form of temperature dependency for surface self-diffusion on copper. They exhibit a low-activation-energy region, $Q_s = 17$ kcal/g-atom in the temperature range 600–850°C, with a tendency toward higher values of Q_s (coupled with higher values of D_s) at higher temperatures. The results of Geguzin et al. [26] on surface self-diffusion on gold by tracer methods and by grooving and smoothing methods show most clearly the increase in effective activation energy with increasing temperature.

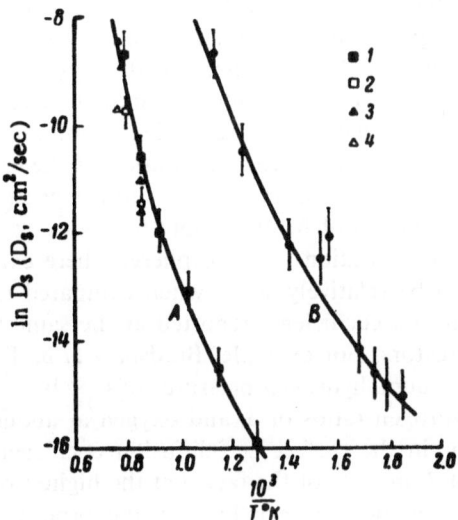

Fig. 1. Temperature dependence of D_s. Results on Curve A were obtained by the mass-transfer method: (1) Scratch healing (data of Geguzin *et al.*). (2) Thermal groove formation (data of Geguzin *et al.*). (3) Thermal groove formation (data of Blakely [27]). (4) Scratch healing (data of Blakely [27]). Results on Curve B were obtained by the layer-specimen method. Taken from Geguzin *et al.* [26].

These results, then, tend to support the idea of a succession of mechanisms controlling the surface-diffusion process as the temperature changes. If a vacancy mechanism, for instance, were to predominate at low temperatures and, possibly, a divacancy mechanism predominate at even lower temperatures, with a rolling-stone mechanism at the highest temperatures, there would be adequate flexibility to explain substantially all the present results, including the upward trend of the apparent activation energy with increasing temperature. However, this is speculation; there is not really enough agreement and sureness about the measurements to support the inference.

Surface contamination from either the bulk or the gas phase may strongly affect the material transport. The existence of a binding energy between impurity atom and point defect could yield an apparent temperature dependence of activation energy. Because they are usually done at the lower temperatures, tracer measurements ought to show specific binding effects more than other methods.

Some investigators describe meticulous vacuum technique, but ignore the possibility that the substrate can supply more than enough impurities to keep the surface contaminated unless it is dynamically cleaned. Sundquist's [29] experiences with equilibrium shapes of particles determined by surface tension should adequately illustrate this point. He found that impurities tended to minimize the variation of the specific surface energy with orientation of the crystal face, and, in fact, this would tend to make the surface energy more nearly isotropic and the model of Mullins more nearly exact.

The effects of variation in atmosphere, where examined systematically, seem to be relatively small when compared to differences in surface self-diffusion coefficients reported at the same temperature by different investigators. For example, Bradshaw et al. [25] find that the conditions of vacuum, hydrogen pressures of 6×10^{-8} atm and 1 atm, water vapor/hydrogen ratios of 5, and oxygen at about 6×10^{-8} atm cause variations in the surface self-diffusion coefficient of copper of only a factor of 7, in spite of the fact that the highest oxygen pressure caused some oxidation, especially near the groove on which the measurements were made.

Unpublished experiments by Drew and Pye [30] on silver surface self-diffusion in partial pressures of oxygen up to 1 atm (too low to form oxide) showed a maximum effect approaching one order of magnitude at pressures below 10^{-6} atm. One of the difficulties in explaining these effects involves showing why they are so small. It appears more difficult to reconcile a rolling-stone mechanism rather than a surface-vacancy mechanism with the small effect of a surface oxide film, unless Bradshaw's observation proves to be a coincidental one applicable only to copper. It will be interesting to see what the surface mobilities look like for other metals with superficial films on them, that is, not just adsorbed impurities, but other phases.

Changes in crystallographic orientation and surface morphology may alter the mobilities of surface atoms to a greater extent than previously conceded. Field-emission observations on tungsten were attributed in part to such large differences. Other field-emission observations for the spreading of foreign atoms on tips indicate very rapid motion with strong directional preference. However, it is possible for some binary systems to depend predominantly on a rolling-stone mechanism (especially where mutual solubility is small) without implying at all that this mechanism is important for surface self-diffusion in either of the elements taken separately.

The near absence of anisotropy of surfaces in the range (110) to (112) reported on copper by Bradshaw *et al.* and the nearly identical behavior of (100) and (110) surfaces [with only a small difference on (111)] of nickel found by Pye and Drew [28] make it seem very unlikely that anisotropic effects will be large enough to explain the many apparent discrepancies in the literature. Although the observations of Blakely and Mykura [10] on nickel are at some odds with those of Pye and Drew, they are attributed in part to impurity effects. The small anisotropy effects seem to be more easily reconciled to a vacancy mechanism than to a rolling-stone model, especially if a long mean free path is included in that model.

However, the measurements published to date have not been accompanied by detailed descriptions of such significant features as faceting, ledge densities, and densities of dislocations emerging through the active surface regions. These aspects of the structure may influence the measurements to a degree, although it is hard to see how they can be used to explain all of the outstanding problems.

In all the theories of planar mobility, including the grain boundary problem, which involve the use of tracers, a characteristic thickness enters, which is normally set equal to a few atomic diameters. However, several sets of observations suggest that the thickness may be greater under some conditions. These reasons may derive more from dynamic than equilibrium sources. For example, Ainslie and Turnbull [31] found that dissolution and diffusion of sulfur into iron grain boundaries generated a dislocation network that increased the sulfur capacity of the expanded grain boundary region far beyond the equilibrium value. Westbrook and co-workers [32] find oxygen effects in AgMg, NiAl, and NiGa grain boundaries that appear to be associated with vacancy-complex formation in micron-width regions. Lundy and co-workers [33]* ascribe abnormal tracer penetration curves in titanium and other metals to a vacancy-depleted region near these metallic surfaces. It is difficult to imagine an equilibrium mechanism that would lead to a depression of the vacancy concentration just below the surface.

Geguzin and co-workers [15,16] cite evidence that their surfaces have an appreciable effective thickness for self-diffusion. Drew and Pye's [24] silver surface self-diffusion measurements as analyzed by the trapping equations of Williams [23] also suggest a surface thickness greatly

---

* In a more recent report [*Acta Met.* **13**: 345 (1965)], Pawel and Lundy withdraw the mechanism proposed previously for this effect. The vacancy-depleted zone no longer appears to be necessary to explain their diffusion results.

exceeding a few atom layers. None of these results appears to be conclusive yet. However, the limitations of both theories and experiments should be kept in mind constantly until truly critical tests can be devised.

It remains possible that the mobility component parallel to the surface varies with depth in a continuous fashion so that the thickness parameter is only an effective value that makes the differential equation fit the problem reasonably well. Furthermore, this thickness may be a function of temperature and of impurity concentration as far as present evidence goes.

Only with the tracer method is trapping into the volume an observable phenomenon. In view of the uncertainty of the surface thickness, trapping can probably be defined most satisfactorily in terms of a decrease in the mobility component parallel to the surface. The observation of trapping must depend on the resolution of the method for detecting spreading. If autoradiography is employed, the critical factor is the range of the radiation in the emulsion. As a rough average, this range might be taken as 20 μ, or about 10^5 times as large as the atomic diameter, the minimum jump distance for a unit diffusion step. Practical measurements require spreading of at least an order of magnitude greater than the minimum resolvable difference. Therefore, to observe surface diffusion at all, the atom must take at least 10^6 net jumps in one direction y parallel to the surface before it takes enough net jumps normal to the surface to penetrate a distance δ. However,

$$\bar{y}^2 = 2D_s t \tag{1}$$

and

$$\delta^2 = 2D_v t \tag{2}$$

If it is assumed that $(\bar{y}/\delta)^2 = \bar{y}^2/\delta^2$,

$$\frac{\bar{y}}{\delta} = \sqrt{\frac{D_s}{D_v}} \tag{3}$$

For no observable interference due to trapping and for $\delta = a_0$, its minimum value, the minimum possible value of $\bar{y}/\delta$ is 10^6. Therefore, trapping may affect the measurements for short times unless the following condition holds:

$$D_s \geqslant 10^{12} D_v \tag{4}$$

For the same minimum δ and greater spreading, trapping might affect the measurements at even larger D_s/D_v ratios. If δ is 100 jump distances, the critical ratio must obey the following relation:

$$\frac{D_s}{D_v} \geqslant 10^8 \tag{5}$$

These are fairly stringent conditions. For mean values of the spreading, $\bar{y}$, greater than 10^6 times the lattice parameter, the critical ratio must be even greater than the values given here.

The effect of trapping should be to reduce the mobility parallel to the surface to a limiting value that would eventually equal that in the bulk of the material. Furthermore, the effective value of the surface self-diffusion coefficient, as measured without allowing for this effect in the calculation, should pass through a maximum value before descending to this limit of the volume self-diffusion coefficient with increasing temperature. Thus, there is an internal way to observe whether the experimental measurements measure the surface or volume diffusion coefficient.

Only the work of Drew and Pye on silver shows the expected trapping effects, and these are shown under conditions where the mean penetration into the surface is fairly large. Their nickel measurements do not exhibit trapping at the same ratios of surface to volume self-diffusion. All of Shewmon's [34] tracer measurements are under conditions of relatively large volume penetration, and some of the other tracer measurements also were performed under these conditions. If the trapping had not been taken into account, the apparent diffusion coefficients ought to be bigger than those reported. The values at the lower temperatures would be raised relative to the extrapolation of the high-temperature values, and it would appear that the change from high activation energy to low activation energy would be even further accentuated.

Finally, if some mechanism inhibits the exchange of an atom on the surface with other atoms below the surface, trapping would be less effective than assumed here; and the surface thickness would be over-estimated. These trapping mechanisms, if they exist, would presumably have to be of chemical origin. There are numerous evidences in volume self-diffusion measurements of failure to transfer tracer from sources placed on the surface into the volume, the reasons for which are generally obscure. We must not ignore this experience until we understand it better.

CONCLUSION

Therefore, we feel that the problem of surface self-diffusion is not completely solved and that there are many details that require a great deal of careful attention. We are no more confident than others before us that our measurements, when completed, will have solved the problem more fully than we think it is now solved.

ACKNOWLEDGMENTS

This work is sponsored by the Air Force Office of Scientific Research of the Office of Aerospace Research under Contract No. AF 49 (638)-1343.

REFERENCES

1. N. A. Gjostein, *Trans. Met. Soc. AIME* **221**: 1039 (1961).
2. P. G. Shewmon and J. Y. Choi, *Trans. Met. Soc. AIME* **227**: 515 (1963); J. Y. Choi and P. G. Shewmon, *Trans. Met. Soc. AIME* **224**: 589 (1962).
3. J. P. Hirth and K. L. Moazed, "Nucleation Processes in Deposition onto Substrates," in: *Fundamental Phenomena in the Materials Sciences*, Vol. 3, Plenum Press (New York), 1966, pp. 63–84.
4. C. E. Birchenall, *Trans. Met. Soc. AIME* **227**: 784 (1963).
5. E. W. Müller, in: *Structure and Properties of Thin Films*, C. A. Neugebauer, J. B. Newkirk, and D. A. Vermilyea (eds.), John Wiley & Sons (New York), 1959, p. 476.
6. R. O. Simmons and R. W. Balluffi, *Phys. Rev.* **117**: 52 (1960); **119**: 600 (1960).
7. G. C. Kuczynski, *Trans. AIME* **185**: 169 (1949).
8. R. Udin, A. J. Shaler, and J. Wulff, *Trans. AIME* **185**: 186 (1949); H. Udin, *Trans. AIME* **189**: 63 (1951).
9. W. W. Mullins, *J. Appl. Phys.* **28**: 333 (1957); **30**: 77 (1959); *Trans. AIME* **218**: 354 (1960).
10. J. M. Blakely and H. Mykura, *Acta Met.* **9**: 23 (1961).
11. J. P. Barbour, F. M. Charbonnier, W. W. Dolan, W. P. Dyke, E. E. Martin, and J. K. Trolan, *Phys. Rev.* **117**: 1452 (1960).
12. R. A. Nickerson and E. R. Parker, *Trans. ASM* **42**: 376 (1950).
13. W. C. Winegard and B. Chalmers, *Can. J. Phys.* **30**: 422 (1952).
14. N. Hackerman and N. Simpson, *Trans. Faraday Soc.* **52**: 628 (1956).
15. Ya. E. Geguzin and G. N. Kovalev, *Soviet Phys. Solid State* (*English Transl.*) **5**: 1227 (1963).
16. Ya. E. Geguzin, G. N. Kovalev, and A. M. Ratner, *Phys. Metals Metallog.* (*USSR*) (*English Transl.*) **10**: 45 (1960).
17. J. C. Fisher, *J. Appl. Phys.* **22**: 74 (1951).
18. R. T. P. Whipple, *Phil. Mag.* **45**: 1225 (1954).
19. P. V. Pavlov and V. A. Panteleev, *Soviet Phys. Solid State* (*English Transl.*) **6**: 955 (1964).
20. J. M. Blakely, *Progr. Mater. Sci.* **10**: 395 (1963).

21. P. G. Shewmon, *J. Appl. Phys.* **34**: 755 (1963).
22. H. Amar and J. B. Drew, *J. Appl. Phys.* **35**: 533 (1964).
23. J. M. Williams, unpublished research.
24. J. B. Drew and J. J. Pye, *Trans. Met. Soc. AIME* **227**: 99 (1963).
25. F. J. Bradshaw, R. H. Brandon, and C. Wheeler, *Acta Met.* **12**: 1057 (1964).
26. Ya. E. Geguzin, G. N. Kovalev, and N. N. Ovcharenko, *Soviet Phys. Solid State (English Transl.)* **5**: 2627 (1964).
27. J. M. Blakely, *Trans. Faraday Soc.* **57**: 1164 (1961).
28. J. J. Pye and J. B. Drew, *Trans. Met. Soc. AIME* **230**: 1500 (1964).
29. B. E. Sundquist, *Acta Met.* **12**: 67 and 585 (1964).
30. J. B. Drew and J. J. Pye, private communication.
31. N. G. Ainslie, V. A. Phillips, and D. Turnbull, *Acta Met.* **8**: 528 (1960).
32. A. U. Seybolt, J. H. Westbrook, and D. Turnbull, *Acta Met.* **12**: 1456 (1964).
33. T. S. Lundy and T. L. Boswell and also R. E. Pawel and T. S. Lundy, presented at AIME, N. Y. meeting, Feb. 1964.
34. J. Y. Choi and P. G. Shewmon, *Trans. Met. Soc. AIME* **230**: 123 (1964).
35. G. E. Rhead, *Acta Met.* **11**: 1035 (1963).
36. C. H. Li and E. R. Parker, unpublished research.
37. A. J. Melmed and R. Gomer, *J. Chem. Phys.* **34**: 1802 (1961).
38. E. Menzel, *Z. Physik* **132**: 508 (1952).
39. Ya. E. Geguzin and N. N. Ovcharenko, *Soviet Phys. Doklady (English Transl.)* **5**: 155 (1960).
40. J. M. Blakely and H. Mykura, *Acta Met.* **11**: 399 (1963).
41. J. M. Blakely and H. Mykura, *Acta Met.* **10**: 565 (1962).
42. P. C. Bettler and F. M. Charbonnier, *Phys. Rev.* **119**: 85 (1960).
43. Kh. Noimann, E. Kloze, and I. L. Sokolskaya, *Soviet Phys. Solid State (English Transl.)* **6**: 1369 (1964).

Defects Near Ionic Crystal Surfaces

Che-Yu Li and J. M. Blakely

Department of Materials Science and Engineering
Cornell University
Ithaca, New York

INTRODUCTION

Changes in the defect distribution on approaching the free surface of a crystal may be of considerable importance in understanding many surface phenomena. A detailed knowledge of the atomic arrangement in the surface region is not readily obtainable from experiment, but it is of interest to investigate from a theoretical viewpoint the influence that point defects are likely to have on certain surface properties. We will discuss here and review some attempts to predict the distribution of point defects in the vicinity of ionic crystal surfaces and examine how some of these ideas may be applied in the treatment of morphological changes governed by capillarity. For simplicity, we will limit the discussion to ionic crystals having the NaCl structure and in which only defects of the Schottky type occur.

It was recognized by Frenkel [1] that, for a crystal in which the values of the work to remove positive and negative ions from the interior to form vacancies were unequal, the condition of zero space charge within the crystal could be satisfied, provided a difference in electrostatic potential existed between the interior of the crystal and possible vacancy sources. Thus, for the process of formation of vacancies in the bulk from the crystal surface, the potential difference V_B between the bulk and surface is of such a magnitude and sign that the electrostatic potential energy difference of the two defects exactly cancels the difference in their separate works of formation. Corresponding to the potential variation between bulk and surface, a region of nonzero space charge will exist in which the concentration of vacancies of one type will be decreased, while that of the other type will be increased. The nature of the surface region has been discussed by a number of authors, e.g., Frenkel [1], Grimley [2], Lehovec [3], and Allnatt [4]. Allnatt has considered the case of a crystal which may contain impurity ions as an additional defect and he has derived general expressions in terms of thermodynamic quantities for the defect chemical potentials, including defect–defect interactions, at any point in the crystal. The special cases treated in detail by Frenkel [1] and Lehovec [3] are best understood in terms of this more general formalism. In the following section, some of the principal features of this work will be described. The last two sections will be concerned with the formation energy of Schottky pairs at a (100) NaCl surface and with morphological changes governed by capillarity.

DEFECT DISTRIBUTION NEAR AN IONIC CRYSTAL SURFACE

Following Allnatt [4], we will consider the equilibrium of a uni-univalent salt MY containing a substitutional divalent metallic impurity I. The subscripts $+$, $-$, and I refer to cation vacancies, anion vacancies, and impurity ions, respectively. If n_i is the number of species of type i per unit volume, then the chemical potential μ_i of that species may be defined by the following relation:

$$\mu_i = \left(\frac{\partial F}{\partial n_i}\right)_{T,v,n_j} \qquad \text{for} \quad j \neq i \tag{1}$$

where F is the Helmholtz free energy per unit volume of the crystal, T is the temperature, and v is the volume. In terms of the free energy change ΔF_i^0 on forming the defect i, where ΔF_i^0 does not include interaction with the space-charge field, interaction with other defects, and the entropy of mixing, the expression for the defect chemical potential at any position is

$$\mu_i = \Delta F_i^0 + q_i V + kT \ln C_i \gamma_i \tag{2}$$

where V is the electrostatic potential* at the site of the defect, γ_i is the local value of the activity coefficient, q_i is the effective charge of the defect, and C_i is the local concentration of defect i, with

$$C_i = \frac{n_i}{z_i}$$

* The reference zero potential is taken at the surface of the crystal.

where n_i is the local number of species i per unit volume and z_i is the number of sites per unit volume available to species i. The evaluation of the activity coefficient requires a detailed statistical-mechanical approach [4], as well as a knowledge of the interaction potential between any pair of defects at any position in the crystal. For cation and anion vacancies, we have

$$\mu_+ = \Delta F_+{}^0 - |e|V + kT \ln C_+ \gamma_+ \tag{3a}$$

$$\mu_- = \Delta F_-{}^0 + |e|V + kT \ln C_- \gamma_- \tag{3b}$$

At equilibrium,

$$\mu_+ = \mu_- = 0 \tag{4}$$

so that

$$C_+ \gamma_+ = \exp\left(-\frac{\Delta F_+{}^0 - |e|V}{kT}\right) \tag{5a}$$

$$C_- \gamma_- = \exp\left(-\frac{\Delta F_-{}^0 + |e|V}{kT}\right) \tag{5b}$$

These equations are valid at any position in the crystal. In particular, in the bulk at large distances from the surface where $V = V_B$, we obtain the following relation from equations (5a) and (5b):

$$V_B = \frac{1}{2|e|}\left(\Delta F_{+B}^0 - \Delta F_{-B}^0 + kT \ln \frac{C_{+B}\gamma_{+B}}{C_{-B}\gamma_{-B}}\right) \tag{6}$$

For future use, we note the following:

$$C_{+S} = C_{+B}\frac{\gamma_{+B}}{\gamma_{+S}}\exp\left(\frac{\Delta F_{+B}^0 - \Delta F_{+S}^0 - |e|V_B}{kT}\right) \tag{7}$$

where the subscripts B and S refer to the bulk and surface, respectively; a similar expression can be applied to C_{-S}.

Similarly, by using an equation of the form of equations (3a) and (3b) for the impurity ions and noting that at equilibrium μ_I is constant throughout the crystal, i.e.,

$$\mu_I = \mu_{IB} \tag{8}$$

we find

$$C_I = C_{IB}\frac{\gamma_{IB}}{\gamma_I}\exp\left(\frac{\Delta F_{IB}^0 - \Delta F_I^0 + |e|V_B - |e|V}{kT}\right) \tag{9}$$

Equations (5a) and (5b) for C_+ and C_- can be written in a form similar to equation (9). The distribution of vacancies and impurities near the surface of the crystal can thus, in principle, be found if the potential variation can be established. The potential V satisfies Poisson's equation:

$$\frac{d^2V}{dy^2} = \frac{-4\pi\rho_c}{\kappa} \tag{10}$$

where y is the coordinate normal to the surface with $y = 0$ at the surface. If z is the number of cation or anion sites per unit volume, the space-charge density ρ_c is given by the following relation:

$$\rho_c = |e|(-C_+ + C_- + C_I)z \tag{11}$$

Lehovec [3] has considered the special case of a pure crystal and has neglected the defect–defect interactions, i.e., γ's $= 1$. Since, in this instance, $C_{+B} = C_{-B}$, equation (6) becomes

$$V_B = \frac{1}{2|e|}(\Delta F^0_{+B} - \Delta F^0_{-B}) \tag{12}$$

and, with the further assumption that the entropies of formation of the positive- and negative-ion vacancies are equal, we obtain

$$V_B = \frac{1}{2|e|}(\Delta E^0_{+B} - \Delta E^0_{-B}) \tag{13}$$

where the ΔE's are the corresponding increases in internal energy for forming the defects. Use of the calculated values of Mott and Littleton [5] yields $V_B \cong -0.28$ V in the case of NaCl.* An additional assumption is made in Lehovec's solution of the potential problem. It is implicitly assumed that the internal energies of formation of the vacancies are independent of position in the crystal. This assumption will be considered in some detail in the next section, where the calculation of vacancy formation energies at an ideal (100) NaCl surface will be outlined.

---

* In obtaining this value of V_B for NaCl, we have taken $W_- - W_+ = \Delta E^0_{-B} - \Delta E^0_{+B}$. This involves the assumption that E_+ and E_- are separately definable quantities [1] and, furthermore, that the energies regained on placing a positive and negative ion back on the crystal are equal.

Using the following boundary conditions:

$$\frac{dV}{dy} = 0 \qquad \text{at} \quad y = \infty$$

$$V = 0 \qquad \text{at} \quad y = 0$$

$$V = V_B \qquad \text{at} \quad y = \infty$$

and the assumptions mentioned above, Lehovec obtains the following expression relating the potential V and distance y:

$$\ln\left[\frac{\exp\left(+\frac{|e|V_B - |e|V}{2kT}\right) + 1}{\exp\left(+\frac{|e|V_B - |e|V}{2kT}\right) - 1}\right] = \sqrt{2}\frac{y}{\lambda} + \ln\left[\frac{\exp\left(\frac{|e|V_B}{2kT}\right) + 1}{\exp\left(\frac{|e|V_B}{2kT}\right) - 1}\right]$$

(14)

where

$$\lambda = \left[\frac{\kappa kT}{4\pi e^2}\frac{1}{(C_{+B}z)}\right]^{1/2}$$

The distribution of defects may be computed from this equation using $\gamma_+ = \gamma_- = 1$ in equations (5a) and (5b).

In the surface region of a NaCl crystal, a negative space charge should exist corresponding to an increase in the positive-ion vacancy concentration relative to the bulk and a decrease in concentration of anion vacancies. The total excess negative charge within the space-charge region per unit area of surface [3] is given by the following relation:

$$Q = \frac{\kappa}{4\pi}\left[\left(\frac{dV}{dy}\right)_{y=0} - \left(\frac{dV}{dy}\right)_{y=\infty}\right] = \left(\frac{\kappa kT}{\lambda 4\pi}\right)\sqrt{8}\sinh\left(\frac{|e|V_B}{2kT}\right)$$

(15)

For NaCl at 600°K, $Q \sim 1.6 \times 10^{11}$ electronic charges/cm². Since this excess charge must be balanced by an equal and opposite charge on the surface layer, the surface layer must contain a deficiency of negative ions or an excess of adsorbed positive ions. Using the above value for the total charge, the excess concentration of negative-ion vacancies or adsorbed positive ions is thus $\sim 10^{-4}$. The exact way that this charge is distributed is unknown, but the distribution will be that corresponding to the minimum in free energy. The parameter λ in equations (14) and (15) is a measure of the thickness of the space-charge region. At 600°K,

$\lambda \sim 2 \times 10^{-5}$ cm. For ionic transport in very small crystals or in discussions of mass transport in the immediate surface region, the existence of the space charge must be recognized. This will be discussed in detail in a later section.

FORMATION ENERGY OF A VACANCY PAIR AT A (100) NaCl SURFACE

From equations (5a) and (5b), we see that, in order to calculate C_+ and C_-, a knowledge of the variation of the free energies of formation of the vacancies as a function of distance from the surface is required. In Lehovec's treatment, the formation energies are treated as constant. We wish to examine this assumption in some detail and will discuss the limiting case when the vacancies are created on an ideal (100) surface of a NaCl crystal at 0°K. The internal energies of formation for this situation should then be compared with the values for the crystal interior.

The approach used here is similar, in principle, to that used by Mott and Littleton [5] in their calculations of defect energies in the bulk. A number of assumptions are made in the present analysis; the result obtained represents a first-order approximation. The details of the calculation are given elsewhere [6]. The analysis refers to the formation of a vacancy by removing a surface cation, the anion case being exactly analogous.

We shall assume throughout that in the absence of surface defects there is no lattice distortion in the surface region of the type considered by Benson et al. [7]. Thus, we are considering the energy change on forming the defect at an idealized surface rather than at a real surface having some degree of displacement. Consider the cation situated at site i in Fig. 1 at the (100) surface of a semi-infinite NaCl crystal. The binding energy of this ion is

$$\sum_j \varphi_{ij} \quad \text{for} \quad j \neq i \tag{16}$$

where φ_{ij} is the interaction potential of ions i and j, and the summation is taken over all ions of the crystal. We shall take $\varphi_{ij} = \varphi_{ij}^{\mathrm{coul}} + \varphi_{ij}^{\mathrm{rep}}$, where $\varphi_{ij}^{\mathrm{coul}}$ is the coulomb potential and $\varphi_{ij}^{\mathrm{rep}}$ is the repulsive potential. These quantities may be expressed as follows:

$$\varphi_{ij}^{\mathrm{coul}} = \pm \frac{e^2}{r_{ij}} \tag{17a}$$

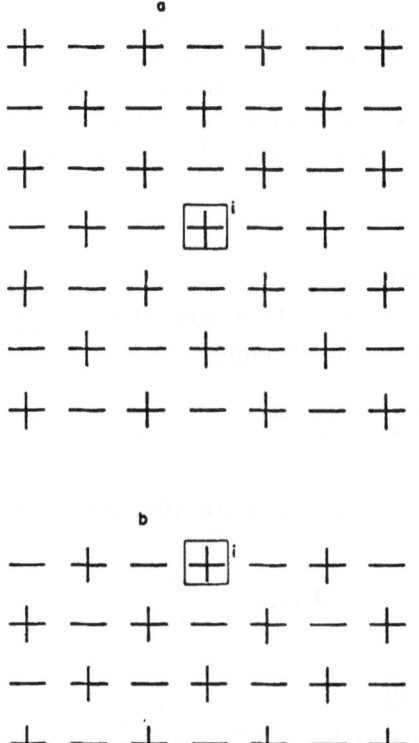

Fig. 1. Schematic of the (100) surface of a semi-infinite NaCl crystal. (a) Top view. (b) Side view.

and

$$\varphi_{ij}^{\text{rep}} = A \exp\left(-\frac{r_{ij}}{\rho}\right) \tag{17b}$$

The work W_+ to remove the cation to infinity from position i is the sum of a number of terms, *viz.*,

W_+ = (Increase in coulomb interaction energy of the crystal) + (Polarization energy of the crystal) + (Increase in repulsive energy of the crystal)

or

$$W_+ = W_{\text{coul}} + W_{\text{pol}} + W_{\text{rep}} \tag{18}$$

The polarization energy is equal to $-\frac{1}{2}eV$, where V is the potential at the site of the missing ion due to the dipoles induced on all other ions of the crystal by the effective charge $-e$ at the vacant site. It may be noted that $\frac{1}{2}eV$ is the work that must be done to depolarize the crystal on replacing the positive ion in its original position, i.e., $W_{pol} = -\frac{1}{2}eV$.

We shall consider each of the contributions in equation (18) separately.

Coulomb Interaction

The change in the coulomb interaction energy of the crystal on removing the positive ion is simply

$$W_{coul} = \sum_j \varphi_{ij}^{coul} = \frac{e^2}{a_0} \sum_j \frac{\pm 1}{p_{ij}} \quad \text{for} \quad j \neq i \tag{19}$$

where $r_{ij} = a_0 p_{ij}$. For an ion in the (100) surface plane of a NaCl-type crystal,

$$W_{coul} = + \frac{e^2}{a_0} (1.679) \tag{20}$$

Polarization Energy

If we assume that the crystal is a continuum of dielectric constant κ, which is independent of position, the polarization P at any point due to the effective charge e at the interface is

$$P = \frac{er}{2\pi |r|^3} \frac{\kappa - 1}{\kappa + 1} \tag{21}$$

where r is the vector from the vacant site. Following Mott and Littleton's [5] procedure, equation (21) may be used to evaluate the induced dipole moments associated with the positive and negative ions. It is found that

$$\mu_1 = \frac{ea_0^3}{2\pi} \frac{r}{|r|^3} \frac{\kappa - 1}{\kappa + 1} \frac{\alpha_1 + \delta}{\frac{1}{2}(\alpha_1 + \alpha_2) + \delta} = A_1 \frac{ea_0^3 r}{|r|^3} \tag{22}$$

and a similar expression is found for μ_2. The terms α_1 and α_2 are the electronic polarizabilities, where the subscripts 1 and 2 refer to cation and anion, respectively, and δ is the displacement polarizability which may be expressed as follows:

$$\delta = e^2/p$$

where p is the force constant due to overlap forces. Since the force constant p is different for the bulk and surface,

$$p_B = \frac{2(1.7476)e^2}{3a_0^2} \left(\frac{1}{\rho} - \frac{2}{a_0} \right) \tag{23a}$$

and

$$p_S = \frac{2(1.7476)e^2}{3a_0^2} \left(\frac{1}{\rho} - \frac{3}{2a_0} \right) \tag{23b}$$

In the derivation of equations (23a) and (23b), only repulsive interactions between first nearest neighbors have been considered. The displacement polarizabilities are, thus, different for surface and bulk ions, and the corresponding expressions for surface and bulk dipole moments must be modified. Thus, we distinguish μ_{1S}, μ_{2S}, μ_{1B}, and μ_{2B}.

As a first approximation, we assume that all dipoles in the crystal are given by equation (22), with the exception of the nearest neighbors of the missing cation, whose displacements and electronic dipole moments are considered to be unknown. If $\xi_p a_0$ and $\xi_q a_0$ are the displacements for ions in positions p and q, respectively, with corresponding electronic dipole moments μ_p and μ_q (see Fig. 2), and

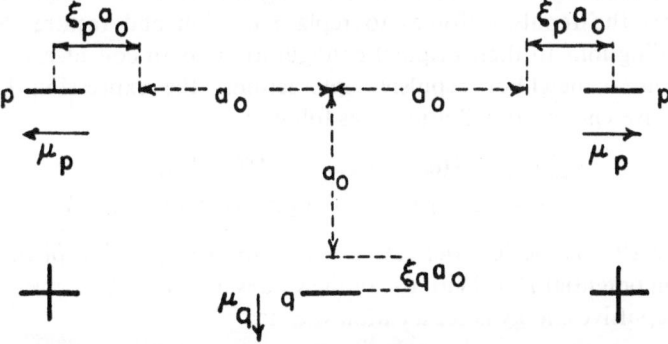

Fig. 2. Schematic diagram showing displacement for ions at positions p and q and their corresponding electronic dipole moments.

if $\mu_p = -m_p e a_0$ and $\mu_q = -m_q e a_0$, we obtain the following relation for the potential at the vacant site:

$$V = \sum_{j}^{(S)} \frac{(\mu_{1S})_j}{r_{ij}^2} + \sum_{j}^{(B)} \frac{(\mu_{1B})_j}{r_{ij}^2} + \sum_{j}^{(S)} \frac{(\mu_{2S})_j}{r_{ij}^2}$$

$$+ \sum_{j}^{(B)} \frac{(\mu_{2B})_j}{r_{ij}^2} + \frac{4e\xi_p}{a_0(1 + \xi_p)^2} + \frac{e\xi_q}{a_0(1 + \xi_q)^2}$$

$$+ \frac{4m_p e}{a_0(1 + \xi_p)^2} + \frac{em_q}{a_0(1 + \xi_q)^2} \tag{24}$$

The subscripts S and B of the summations denote that (excluding the nearest neighbors of the vacancy) the summations are to be taken over surface and bulk ions, respectively. The method of evaluating the ξ's and the m's in equation (24) consists of setting up equations representing the balance of electrostatic and repulsive forces at the nearest-neighbor anion sites. Evaluation of equation (24) for NaCl yields

$$\tfrac{1}{2}eV = \frac{e^2}{a_0}(0.592) \tag{25}$$

·A similar analysis for the case of a missing anion yields

$$\tfrac{1}{2}eV = \frac{e^2}{a_0}(0.547) \tag{26}$$

Repulsive Energy

The contribution of the repulsive forces to the work of extraction of the ion may be found by considering the work that must be done against the repulsive forces to replace the ion and restore the surrounding ions to their original configuration. With consideration only of nearest neighbor repulsive interactions, the expression for the repulsive energy contribution is as follows:

$$4[\tfrac{1}{2}a_0\xi_p\varphi'^{\text{rep}}(a_0 + a_0\xi_p) - \varphi^{\text{rep}}(a_0 + a_0\xi_p)]$$
$$+ \tfrac{1}{2}a_0\xi_q\varphi'^{\text{rep}}(a_0 + a_0\xi_q) - \varphi^{\text{rep}}(a_0 + a_0\xi_q) \tag{27}$$

where φ'^{rep} is the first derivative with respect to position of the interaction potential φ^{rep}. With use of the values of ξ_p and ξ_q as given above, the repulsive energy is for a cation vacancy

$$W_{\text{rep}} = -\frac{e^2}{a_0}(0.125) \tag{28a}$$

and, for an anion vacancy,

$$W_{\rm rep} = -\frac{e^2}{a_0}(0.120) \tag{28b}$$

Finally, we obtain the following expression for the work of extraction:

$$W_+ = \frac{e^2}{a_0}(0.962) \tag{29a}$$

and

$$W_- = \frac{e^2}{a_0}(1.01) \tag{29b}$$

The corresponding results of Mott and Littleton [5] for the creation of positive- and negative-ion vacancies within the bulk are

$$W_+ = \frac{e^2}{a_0}(0.91) \tag{30a}$$

$$W_- = \frac{e^2}{a_0}(1.02) \tag{30b}$$

Thus, within the limits of the assumptions and accuracy of the calculations, we see that the work of extraction of an ion (and, hence, the vacancy formation energy) is approximately the same for the outermost (100) plane as for the crystal interior. It should be noted that, although the final result in the two cases is almost the same, the various contributions to the work of extraction are significantly different.

The variation of the formation energy with distance into the crystal has been considered for the case where induced displacement dipoles are ignored. The results indicate that the formation energy does not vary appreciably with position. These results are shown in the Appendix.

Although a number of assumptions are involved in our treatment (in particular, that the dielectric constant does not vary with position), we may conclude that to a first approximation the vacancy formation energy in NaCl may be taken to be independent of position, as assumed by Lehovec [3].

MORPHOLOGICAL CHANGES IN IONIC CRYSTALS
GOVERNED BY CAPILLARITY

The general theory of mass transport in crystals due to surface tension has been discussed by Herring [8]. Mullins [9] has applied the concepts developed by Herring to the discussion of shape changes in crystals consisting of one atomic species and, in particular, has analyzed the morphological changes that occur in nearly flat surfaces. Since similar changes occur in ionic crystals and these are closely related to sintering and other mass transport phenomena, it is worth considering how the theory of mass transport due to surface tension should be modified for this case. For mass transport in a system consisting of two or more atomic species, the mobilities of the individual species will not, in general, be equal, and the individual fluxes are not independent of each other. This difference in mobilities gives rise to frictional forces [10–13], which may be electrical or chemical in nature. For example, for simple ionic crystals there are fluxes of both positive and negative ions (or vacancies), and the calculation of creep rates requires the inclusion of a frictional force to maintain electrical neutrality [13].

The chemical potential of positive and negative ions at a portion of the surface of curvature K may be written as follows:

$$\mu_1 = \mu_1{}^0 + K\sigma\Omega_1 \tag{31a}$$

$$\mu_2 = \mu_2{}^0 + K\sigma\Omega_2 \tag{31b}$$

where $\mu_1{}^0$ and $\mu_2{}^0$ are values of the chemical potentials corresponding to equilibrium at a flat surface ($K = 0$). In equations (31a) and (31b), we assume that the specific surface free energy σ is independent of the orientation and curvature of the surface. This will be assumed throughout the discussion. Ω_1 and Ω_2 are the volume per cation and anion, respectively.

Consider now a surface of variable curvature. From equations (31a) and (31b), we see that there will be corresponding gradients in chemical potential in the system, and, thus, ions will move in such a way as to make μ_1 and μ_2 independent of position, i.e., the surface will tend to a constant curvature. For an initially nearly planar surface, the surface will become flat. We will consider first the case of mass transport via the mechanism of surface diffusion, and the entire detailed discussion below will be limited to nearly flat surfaces.

The compatibility of the positive- and negative-ion fluxes may be satisfied by including an additional frictional force* $\mathbf{F}$ in the flux equations [10–13] such that

$$\mathbf{J_1}^S = \mathscr{B}_1^S n_1 (\nabla \mu_1 - \mathbf{F}) \tag{32a}$$

$$\mathbf{J_2}^S = \mathscr{B}_2^S n_2 (\nabla \mu_2 + \mathbf{F}) \tag{32b}$$

where n_1 and n_2 are the number of cations and anions per unit area in the outermost layer of the crystal, excluding adsorbed ions, and $\mathscr{B}_1^S$ and $\mathscr{B}_2^S$ are the surface cation and anion mobilities. The net accumulation rates of cations and anions at any point on the surface are $-div\,\mathbf{J_1}^S$ and $-div\,\mathbf{J_2}^S$, respectively. In order that the new element of material thus created maintain the original composition, we must have

$$\frac{div\,\mathbf{J_1}^S}{div\,\mathbf{J_2}^S} = \frac{n_1}{n_2} \tag{33}$$

With the assumption that n_1 and n_2 are slowly varying functions of position on the surface, so that they may be assumed constant, we obtain the following from equations (32a) ,(32b), and (33):

$$div\,\mathbf{F} = \frac{\mathscr{B}_1^S \nabla^2 \mu_1 - \mathscr{B}_2^S \nabla^2 \mu_2}{\mathscr{B}_1^S + \mathscr{B}_2^S} \tag{34}$$

Thus,

$$div\,\mathbf{J_1}^S = \mathscr{B}_1^S n_1 \nabla^2 \mu_1 - \mathscr{B}_1^S n_1 \frac{\mathscr{B}_1^S \nabla^2 \mu_1 - \mathscr{B}_2^S \nabla^2 \mu_2}{\mathscr{B}_1^S + \mathscr{B}_2^S}$$

$$div\,\mathbf{J_1}^S = \frac{n_1 \mathscr{B}_2^S \mathscr{B}_1^S}{\mathscr{B}_2^S + \mathscr{B}_1^S} \nabla^2 (\mu_2 + \mu_1) \tag{35}$$

$$div\,\mathbf{J_1}^S = \frac{n_1 \mathscr{B}_2^S \mathscr{B}_1^S}{\mathscr{B}_2^S + \mathscr{B}_1^S} \sigma \Omega \nabla^2 K$$

where Ω is the volume per anion–cation pair.

For a surface of small slope if the y direction is taken normal to

---

* In calculating creep rates in ionic crystals, Ruoff [13] has represented the frictional force $\mathbf{F}$ as an electrical potential gradient determined by using the condition of electrical neutrality at steady state. However, a frictional force will also exist in any metallic solid solution of species having unequal mobilities. The frictional force, in this case, is necessary to prevent a separation of the components and, hence, a deviation from the steady state. Thus, the frictional force should be represented in terms of a chemical potential gradient.

the mean surface ($y = 0$) and if the only curvature is in the direction x parallel to the mean surface, we obtain

$$\frac{\partial J_1{}^S}{\partial x} = \frac{n_1 \mathscr{B}_2{}^S \mathscr{B}_1{}^S}{\mathscr{B}_2{}^S + \mathscr{B}_1{}^S} \sigma\Omega \frac{\partial^4 y_S}{\partial x^4}$$

where y_S is the y-coordinate of the surface. The rate at which the surface moves normal to itself at any point is

$$\frac{\partial y_S}{\partial t} = \frac{a}{n_1} \frac{\partial J_1}{\partial x}$$

where a is the interplanar spacing normal to the surface. Thus,

$$\frac{\partial y_S}{\partial t} = \frac{D_2{}^S D_1{}^S}{D_2{}^S + D_1{}^S} \frac{a\sigma\Omega}{kT} \frac{\partial^4 y_S}{\partial x^4} = A \frac{\partial^4 y_S}{\partial x^4} \tag{36}$$

where $\mathscr{B} = D/kT$. Thus, equation (36) describes shape changes of a surface of small slope by surface diffusion of both species.

For the case of transport by volume diffusion only, we may again describe the positive- and negative-ion fluxes by equations of the form of equations (32a) and (32b). Thus, for the case of a nearly flat surface, the fluxes normal to the surface at any point are given by

$$(\mathbf{J}_1)_y = \mathscr{B}_1 n_1 \left(\frac{\partial \mu_1}{\partial y} - \mathbf{F}_y \right) \tag{37a}$$

and

$$(\mathbf{J}_2)_y = \mathscr{B}_2 n_2 \left(\frac{\partial \mu_2}{\partial y} + \mathbf{F}_y \right) \tag{37b}$$

where $\mathscr{B}_1$ and $\mathscr{B}_2$ are the mobilities of the ions at distance y below the surface. The assumption is made that the ratio of the concentrations of positive and negative ions at the surface $(n_1/n_2)_{y=0}$ is independent of the local curvature. The ratio of positive- and negative-ion fluxes normal to the surface should be such as to preserve the equilibrium ratio of positive and negative ions. This condition may be written as

$$\left. \frac{(\mathbf{J}_1)_y}{(\mathbf{J}_2)_y} \right|_{y=0} = \left. \frac{n_1}{n_2} \right|_{y=0} \tag{38}$$

From equations (37a), (37b), and (38), we obtain

$$(\mathbf{F}_y)_{y=0} = \left(\frac{\mathscr{B}_1 \dfrac{\partial \mu_1}{\partial y} - \mathscr{B}_2 \dfrac{\partial \mu_2}{\partial y}}{\mathscr{B}_1 + \mathscr{B}_2} \right)_{y=0} \tag{39}$$

Also, the rate at which the surface moves normal to itself is

$$\frac{\partial y_S}{\partial t} = \left[(J_1)_v \frac{a}{n_1}\right]_{y=0}$$

$$\frac{\partial y_S}{\partial t} = \left[\frac{D_1 D_2}{D_1 + D_2} \cdot \frac{a}{kT} \frac{\partial}{\partial y} (\mu_1 + \mu_2)\right]_{y=0} \tag{40}$$

It should be noted that in equation (40) the volume diffusion coefficients D_1 and D_2 are those in the region immediately beneath the surface and that these may differ from the bulk values. If the process of diffusion is by a vacancy mechanism, then the ionic diffusion coefficient D_1 and vacancy diffusion coefficient D_+ are related by the following expression:

$$D_1 = C_+ D_+ \tag{41}$$

where C_+ is the local concentration of positive-ion vacancies. Thus, if D_+ is independent of the local concentration of vacancies, the local value of D_1 may be expressed in terms of the bulk cation diffusion coefficient $D_1{}^B$ by

$$D_1 = \frac{C_+}{C_{+B}} D_1{}^B \tag{42}$$

Using equation (7) with the simplifications noted previously, we obtain from equations (40) and (42)

$$\frac{\partial y_s}{\partial t} = \frac{D_1{}^B D_2{}^B}{D_1{}^B + (e^{+2|e|V_B/kT}) D_2{}^B} \cdot \frac{a}{(e^{-|e|V_B/kT})kT} \cdot \left[\frac{\partial}{\partial y} (\mu_1 + \mu_2)\right]_{y=0} \tag{43}$$

$$\frac{\partial y_s}{\partial t} = C \left[\frac{\partial}{\partial y} (\mu_1 + \mu_2)\right]_{y=0} \tag{44}$$

For NaCl at $\sim 600°$K, $e^{+2|e|V_B/kT} \cong 10^{-5}$ and decreases with increase of temperature.

It should be noted that the rate at which the shape change occurs may be characterized by the quantities A [equation (36)] and C [equation (44)] for the pure surface diffusion and the pure volume diffusion cases, respectively. In either instance, we see that if the diffusion coefficient of one species is sufficiently greater than the other, the mass transport rate is dependent only on the coefficient of diffusion of the more slowly moving species. A similar conclusion has been reached for the creep rate [13] in ionic crystals, neglecting the inhomogeneous space-charge region.

For NaCl at $600°K$ [14], the ratio $D_1^B/D_2^B \approx 25$, so that the quantity in equation (43) that is represented by C in equation (44) becomes

$$C \approx D_2^B \frac{a}{e^{-|e|V_B/kT} \cdot kT} \tag{45}$$

Thus, the transport is controlled by the more slowly moving anion and the factor $(e^{-|e|V_B/kT})^{-1}$ represents the decrease in the local concentration of anion vacancies at the surface due to the existence of the space charge. Thus, in calculating diffusion coefficients and activation energies from experimental values of C, the factor $(e^{-|e|V_B/kT})^{-1}$ must be included. For the surface diffusion case, the mechanism for the motion of the ions is not known. If, however, the motion occurs by a vacancy mechanism or by motion in the adsorbed state, the ion diffusion coefficient may be expressed by an equation similar to equation (41), where C_+ is replaced by the concentration of the moving species of vacancies or adsorbed ions and D_+ is replaced by the diffusion coefficient of that species. The concentration of the vacancies or adsorbed ions will depend on the existence of the space charge. This introduces additional complications in the interpretation of surface diffusion data.

In the derivation of equations (36) and (44), the assumption was made that n_1 and n_2 may be taken to be independent of x and, hence, independent of the curvature K. A similar assumption has been made for the one-component system treated by Mullins [9]. In this way, the concentration gradients do not need to be specified and the problem may be treated in terms of the chemical potentials μ_1 and μ_2. For a more exact treatment of the problem, the variation of n_1 and n_2 with curvature must be considered. This problem has been discussed elsewhere [15].

Finally, solutions of equations of the form of equations (36) and (44) have been obtained by Mullins [9] for two special cases which are of interest in connection with experimental measurements of diffusion coefficients, viz., the development of grooves at surface–grain boundary intersections and the smoothing of single and multiple scratches. These solutions apply in the case of ionic crystals where the quantities A and C derived here replace the corresponding expressions in Mullins' formulation.

Experimental observations of these processes as described previously for metals may thus be used to determine the quantities

Table I

Simplified Calculations for the Three Outermost Layers of NaCl

	Total binding energy $\times\left(\dfrac{a_0}{e^2}\right)$	Polarization energy $\times\left(\dfrac{a_0}{e^2}\right)$		Work to remove ion $\times\left(\dfrac{a_0}{e^2}\right)$	
		Cation	Anion	W_+	W_-
Zero layer	1.501	0.425	0.265	1.076	1.235
First layer	1.564	0.404	0.283	1.160	1.281
Second layer	1.560	0.415	0.285	1.145	1.274
Bulk[5]	1.562	0.430	0.301	1.132	1.261

A or C. These techniques may also be applied to ionic crystals and, in conjunction with ionic conductivity data, should provide useful information concerning the separate diffusion coefficients of the two species.

In conclusion, we have attempted to discuss mass transport in ionic systems where separate cation and anion fluxes must exist and have considered the effect of frictional forces and the space-charge region at the surface. It would be of interest to apply similar ideas to the analysis of sintering phenomena in ionic systems.

APPENDIX: WORK TO REMOVE IONS FROM THE THREE OUTERMOST (100) LAYERS OF NaCl

The first-order type of calculation outlined in the text for the outermost plane requires much numerical calculation. For the sake of comparison, we have used the zero-order approximation as described by Mott and Littleton [5]. The results of these simplified calculations are shown in Table I.

ACKNOWLEDGMENTS

The authors would like to thank Mr. C. A. Steidel for helpful suggestions and Professor A. L. Ruoff for discussions on frictional forces. This work is supported by the Air Office of Scientific Research (Grant No. AF-AFOSR-780-65) and by the U. S. Atomic Energy Commission under contract No. AT(30-1)-3228.

REFERENCES

1. J. Frenkel, *Kinetic Theory of Liquids*, Oxford University Press (London), 1946.
2. T. B. Grimley, *Proc. Roy. Soc. (London)* **A201**: 40 (1950).
3. K. Lehovec, *J. Chem. Phys.* **21**: 1123 (1953).
4. A. R. Allnatt, *J. Phys. Chem.* **68**: 1763 (1964).
5. N. F. Mott and M. J. Littleton, *Trans. Faraday Soc.* **34**: 485 (1938).
6. J. M. Blakely and Che-Yu Li, Report No. 308, Materials Science Center, Cornell University, Ithaca, N. Y., 1965.
7. G. C. Benson, P. I. Freeman, and E. Dempsey, *Solid Surfaces, Advances in Chem. Series, No.* 33, Am. Chem. Soc. (Washington, D. C.), 1961, p. 26.
8. C. Herring, "Surface Tension as a Motivation for Sintering," in: *The Physics of Powder Metallurgy*, McGraw-Hill (New York), 1951.
9. W. W. Mullins, "Morphologies Governed by Capillarity," in: *Metal Surfaces*, ASM, 1962.
10. W. Nernst, *Z. Physik. Chem. (Leipzig)* **2**: 613 (1888).
11. C. Wagner, *Z. Elektrochem.* **65**: 581 (1961).
12. R. W. Laity, *J. Phys. Chem.* **67**: 671 (1963).
13. A. L. Ruoff, Report No. 299, Materials Science Center, Cornell University, Ithaca, N. Y., 1965.
14. D. Patterson, G. S. Rose, and J. A. Morrison, *Phil. Mag.* **46**: 393 (1955).
15. J. M. Blakely and Che-Yu Li, "Changes in Morphology of Ionic Crystals Due to Capillarity," *Acta Met.* **14**: 279 (1966).

Effect of Surfaces on Mechanical Behavior of Metals

I. R. Kramer

Martin Company
Baltimore, Maryland

INTRODUCTION

That the surface has a pronounced effect on the mechanical behavior of metals is well established. In spite of this, surface effects are often ignored with the result that many of the explanations concerning plastic-flow phenomena are inadequate. In the following discussion, it will be shown that surface effects exert a remarkable influence on creep and fatigue, as well as on the general plastic-flow characteristics of metals. The change in mechanical behavior has been studied on specimens with oxide and metallic films, both when subjected to a vacuum or inert-gas environment and when the surface was being removed during the deformation. The mechanical behavior was also found to be influenced by nonpolar solutions containing surface-active molecules.

SOLID FILMS

In 1934, Roscoe [1] found that applying an oxide film less than 20 atoms thick to cadmium crystals caused an increase of 50 % in the initial flow stress. An increase in the thickness of the oxide film to approximately 1200 atoms increased this stress by nearly 100%. When the oxide film was removed by brushing the surface with a dilute acid, the flow stress returned to its normal value. Later, Cottrell and Gibbons [2] confirmed Roscoe's results and reported that the presence of a thin oxide film on cadmium crystals increased the critical resolved shear stress from 12 to 30 g/mm². Harper and Cottrell [3] obtained similar results on zinc crystals which had been oxidized with steam. Takamura [4] determined the behavior of aluminum crystals with different thicknesses of oxide films. According to these results, the

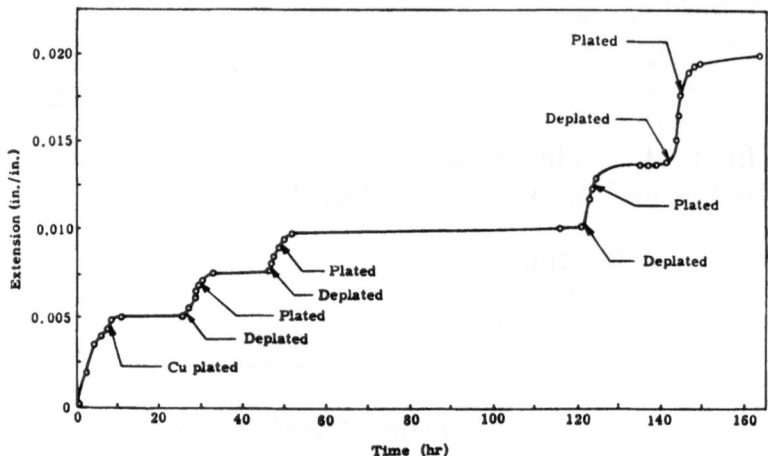

Fig. 1. Change in creep rate of a zinc single crystal produced by a thin copper plate. Taken from Pinkus and Parker [5].

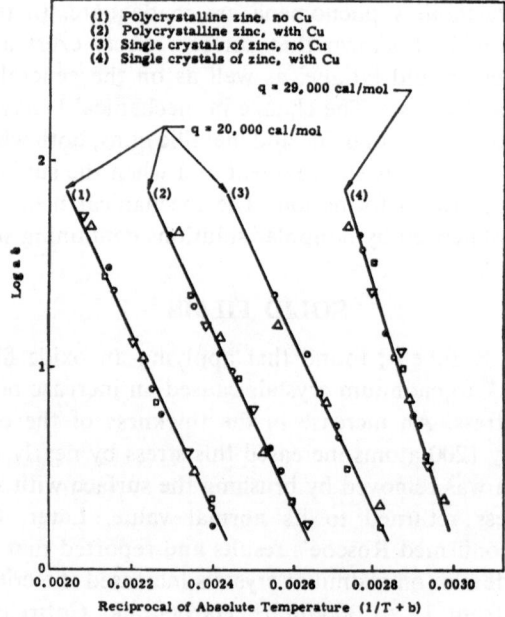

Fig. 2. Composite plot of data from 18 tests designed to establish the temperature dependence of the creep rate for oxide-free polycrystalline and single-crystal zinc both with and without copper on the surface. Taken from Pinkus and Parker [5].

critical resolved shear stress changed from 76 to 174 g/mm² when the oxide thickness was increased from 100 to 500 Å.

The effect of a thin copper plate on the creep behavior of a zinc single crystal is shown in Fig. 1 [5]. In these experiments, the crystal was plated and deplated alternately during the creep process. When the specimen was plated, the creep rate decreased sharply, and when deplated, it assumed the same creep rate as that of the virgin crystal. The activation energy for creep for polycrystalline and single-crystal zinc, with and without the copper plating, is shown in Fig. 2, where the creep rate is designated as $\dot{\epsilon}$ and where a and b are constants. The activation energy of the single-crystal specimens decreased from 29,000 to 20,000 cal/mole when plated, whereas the temperature dependence of the polycrystalline specimens was not affected. In the above experiments, it was shown that on polycrystalline specimens the presence of a thin layer of copper had no effect on the creep behavior, and it was necessary to diffuse the copper into zinc to form an alloy layer in order to modify the mechanical behavior.

VACUUM AND ATMOSPHERIC-GAS EFFECTS

Prior to the work of Roscoe [1], Gough and Sopwith [6] studied the fatigue behavior of metals at a pressure of 10^{-3} torr. The endurance limit of lead was more than doubled, while the endurance limits of copper and brass were increased 13 and 26%, respectively. The reduction of fatigue life was found to be associated with the presence of oxygen and water vapor. Neither gas alone was very effective [7,8]. Wadsworth [9] found that the fatigue lives of copper, aluminum, and gold were increased by factors of 20, 10, and 1, respectively, when the pressure was reduced from atmospheric to 10^{-5} torr. The presence of water vapor at decreased pressure reduced the fatigue life of aluminum, but not that of copper, while oxygen reduced the fatigue life of both copper and aluminum. The relationship between pressure and fatigue life for copper is shown in Fig. 3 [9]. Snowden et al. [10,11], in a study of lead at 5×10^{-3} torr, found that, in addition to an appreciable increase in fatigue life, the surface of the specimens had a much greater number of slip markings and furrows than the surfaces of specimens tested at atmospheric pressures. These observations tend to confirm those of Wadsworth [9], who reported that the increase in fatigue life in vacuum is associated with the rate of propagation of the cracks and not with their initiation. Apparently, cracks can form equally well in vacuum as

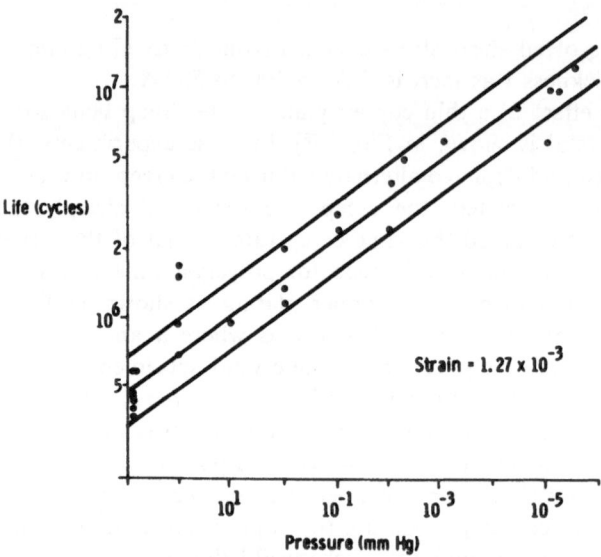

Fig. 3. Relation between fatigue life and air pressure for copper.

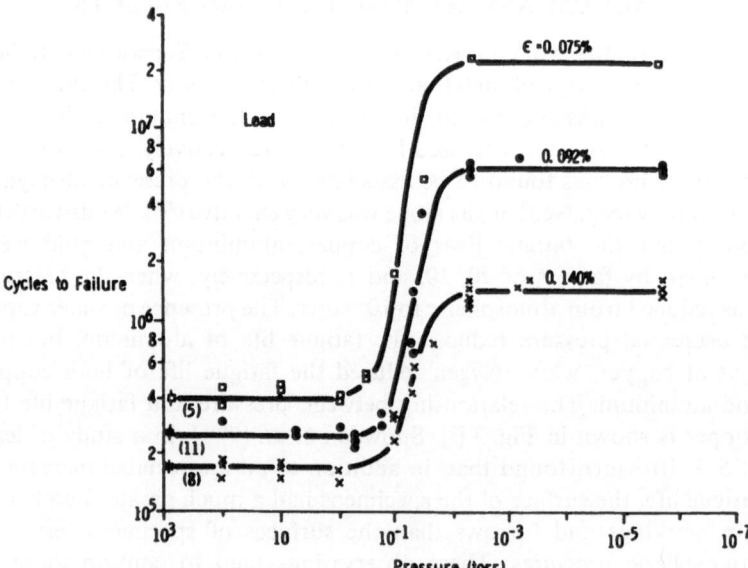

Fig. 4. Variation of fatigue life with air pressure. (The values in parentheses refer to the number of specimens tested at normal atmospheric pressure; all other experimental points are plotted.)

in air, but, in the former case, many cracks never reach the critical size required for propagation.

The fatigue life of lead (Fig. 4) as a function of pressure [11] does not appear to follow the same relationship as that reported for copper [9]. Instead of increasing continuously with decreasing pressure, the fatigue life of lead remained rather constant down to pressures of 10^{-1} torr, whereas it increased rapidly and again remained constant at pressures below about 5×10^{-3} torr. Snowden [12] suggested that this difference in behavior may be due to size and geometry of the cracks and their effect on gas flow due to the crack tip. It is of interest to note in Fig. 4 that the improvement in fatigue life at reduced pressures is a function of the stress or strain amplitude. At a strain of 0.14%, the number of cycles to failure increased from approximately 1.7×10^5 to 1.5×10^6, while at a strain of 0.075%, the life increased from 4×10^5 to 2×10^7 cycles. Kramer et al. [13] found that the fatigue life of commercially pure aluminum tested at reduced pressures was similar to that of lead (Fig. 5). The fatigue life increased very slowly until the pressure was reduced to 3×10^{-2} torr and then increased rapidly as the pressure was reduced further. Below a pressure of about 3×10^{-5} torr, a second plateau region was reached and the fatigue life increased very slowly with further reductions in pressure. In addition to the pressure effects, the increase in fatigue life at reduced pressure was found to be frequency-dependent. As noted in Fig. 5, the fatigue life of specimens tested at 49 cps is higher than that of specimens tested at 76 cps. Achter, Danek, and Smith [14] also reported a plateau region in the fatigue–pressure relationship for nickel tested at 816°C. The pressure at which the fatigue life appeared to become insensitive to further reductions occurred at approximately 10^{-3} torr.

Since the improvment in fatigue life of metals is associated with the absence of reactive species in the atmosphere, it is natural to expect an alteration in mechanical behavior when specimens are tested in an inert atmosphere. Thompson, Wadsworth, and Louat [15] reported that the fatigue life of copper was five times greater in nitrogen than that in air. The introduction of water vapor was reported by Mantel [16] to reduce the fatigue life of a 52100 steel.

Not only does the composition of the atmosphere affect the fatigue life of metal, it also exerts an influence on the creep rate and stress–rupture life. McCoy and Douglas [17] (Fig. 6) found that the creep rate of 304 stainless steel at 1700°F was higher in argon, nitrogen, oxygen, and hydrogen than that in air. Shahanian [18] and Shahanian

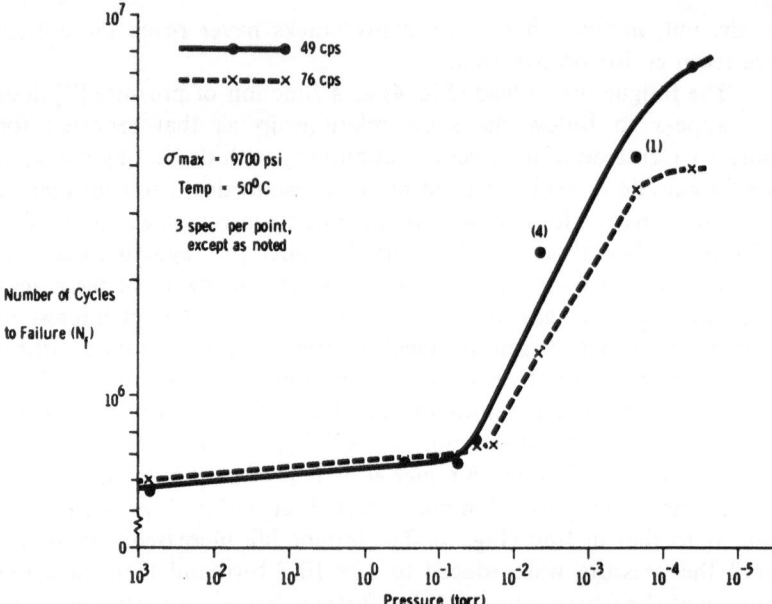

Fig. 5. Variation of fatigue life with pressure at 49 and 76 cps for aluminum 1100.

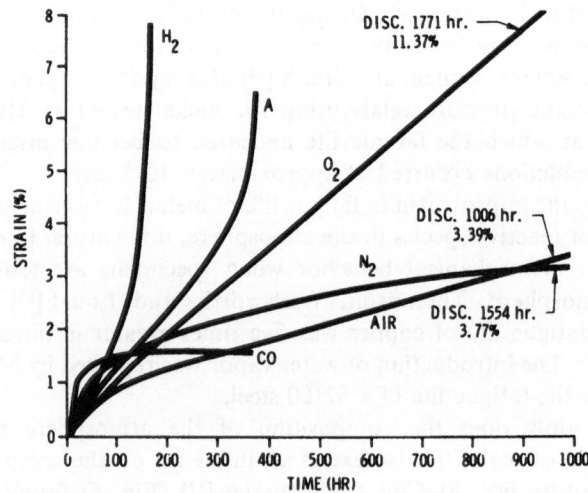

Fig. 6. Effect of environment on the creep properties of type-304 stainless steel at
1700°F and 1200 psi.

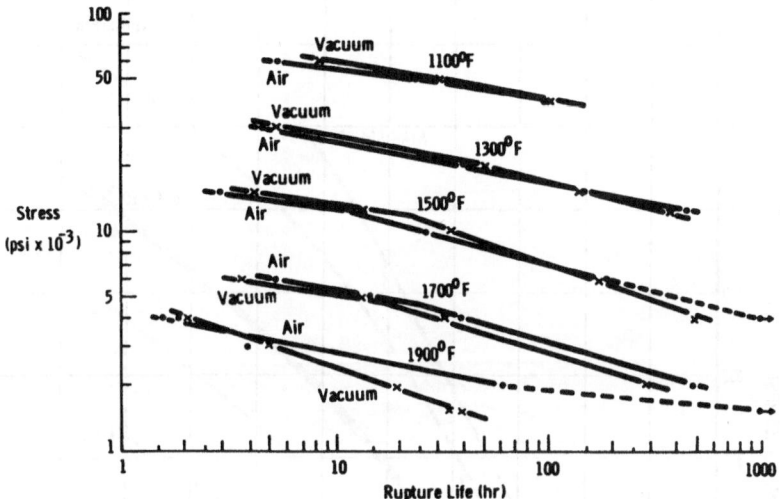

Fig. 7. Influence of temperature and stress on the rupture life of 80 Ni–20 Cr alloy in air and in vacuum.

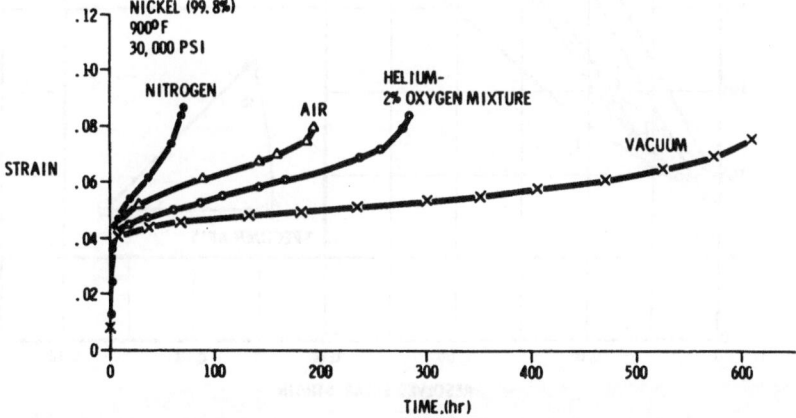

Fig. 8. Creep of nickel at 900°F and 30,000 psi.

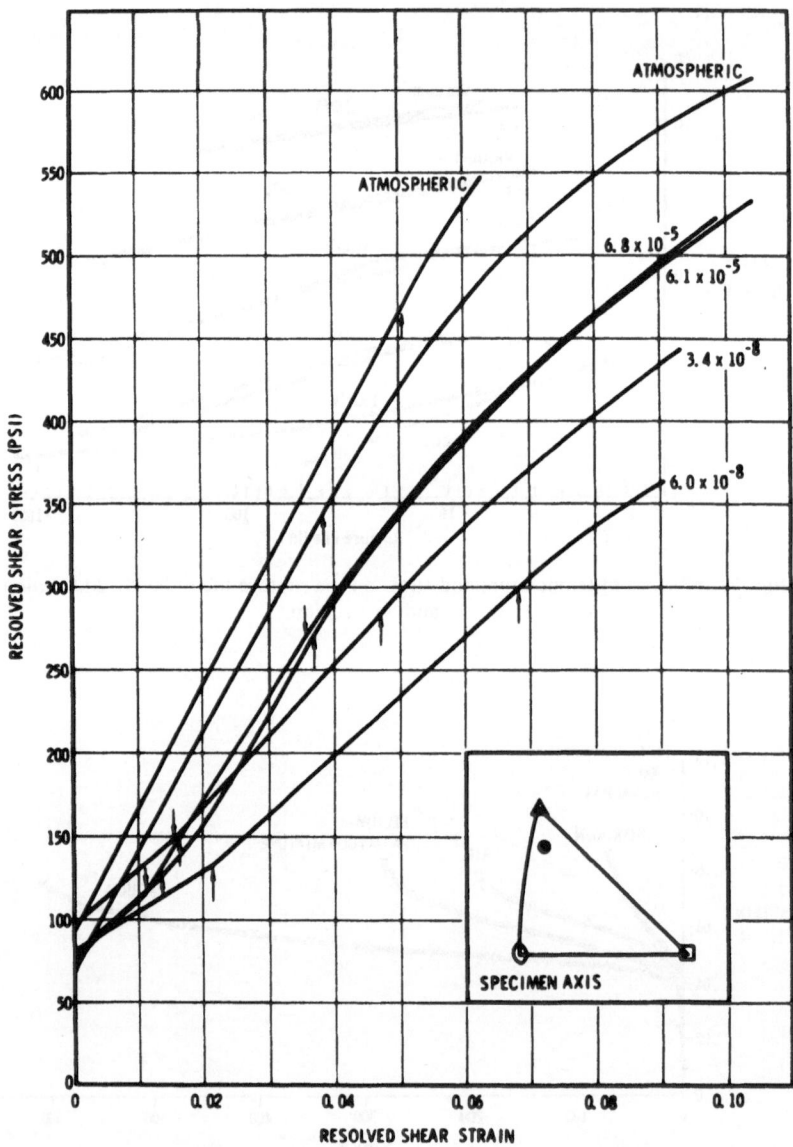

Fig. 9. Stress–strain curves for single-crystal aluminum in air and in vacuum.

and Achter [19] reported that the creep–rupture life and creep resistance of low-alloy steels and various high-temperature alloys were greater in air than in oxygen, nitrogen, helium, and vacuum; however, the rupture life of an 80 Ni–20 Cr alloy was greater in air than in vacuum only at high temperatures and low strain rates. At low temperatures and high strain rates, the opposite behavior occurred; that is, the specimens tested in vacuum had the higher rupture life (Fig. 7). These authors [20] also reported that the creep properties of nickel in air were considerably lower than those in vacuum, and testing in nitrogen caused a further decrease (Fig. 8). To explain this effect, the authors [20] stated that nitrogen reduced the creep strength by reacting with impurities in the grain boundaries and facilitated crack propagation.

While an appreciable effort has been devoted to studies on the effect of oxide and metallic coatings, very little has been done on the change in tensile characteristics of specimens deformed in a vacuum environment. Kramer and Podlaseck [21] found that the various deformation stages were affected when aluminum single crystals were pulled at reduced pressures (Fig. 9). For the crystals used in the investigation, the orientation was such that a Stage I region did not appear in tests conducted at atmospheric pressure. When the pressure was reduced to about 10^{-5} torr, not only did a Stage I region appear,

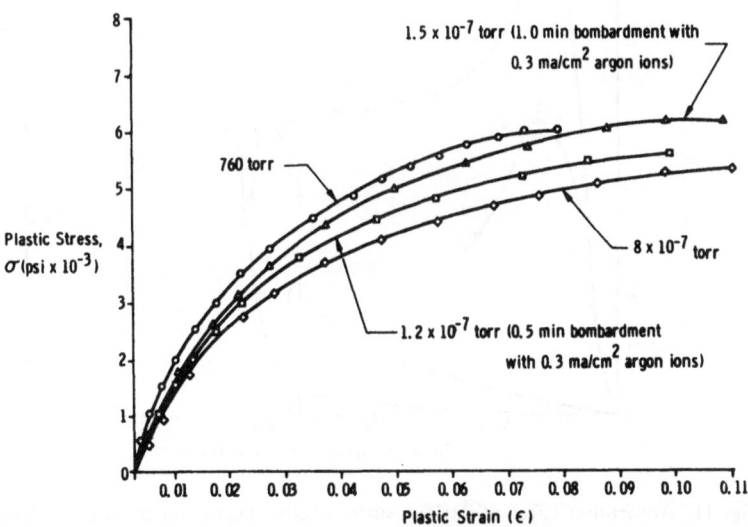

Fig. 10. Plastic-flow curves for high-purity aluminum wire.

but also the slope of the Stage II region decreased. With further reductions in pressure, the extent of Stage I increased and the slopes of both Stages I and II decreased. Figure 10 shows the effect of low pressure on the plastic-flow behavior of annealed, polycrystalline, high-purity aluminum. Wire specimens of 0.020-in. diameter were pulled at a strain rate of 10^{-5} sec^{-1} at a pressure of 8×10^{-7} torr. It is seen that at the higher strains the flow stress is considerably lower than that of specimens pulled in air. In addition, the total elongation to fracture was increased approximately 40%.

SURFACE-ACTIVE AGENTS

It has been known for some time that the mechanical behavior of metals (or minerals) may be changed when tests are conducted in a nonpolar medium containing surface-active agents. Rehbinder [22] and Rehbinder and Wenstrom [23] observed that the creep rate of lead, tin, and copper sheets, under constant load, was much greater if small amounts of surface-active agents (cetyl alcohol and n-valeric, n-heptoic, stearic, oleic, palmitic, and cerotic acids) were added to the paraffin-oil

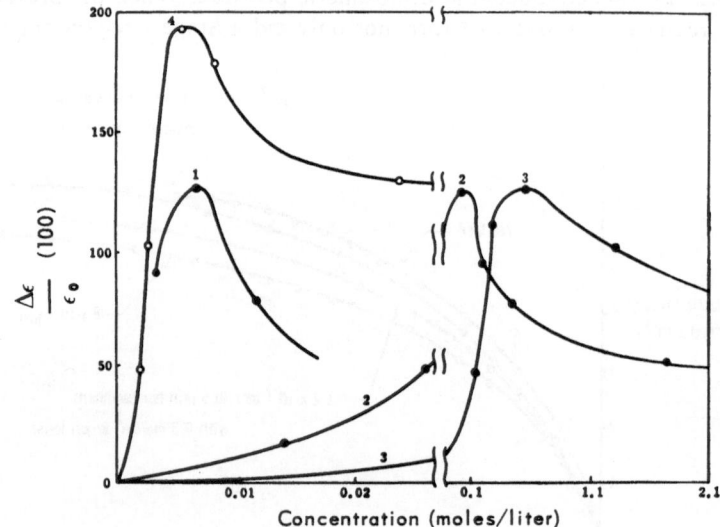

Fig. 11. Adsorption effect of various surface-active agents on the creep behavior of tin single crystals. Solvent, octane. 1—Stearic acid. 2—Caprylic acid. 3—Propionic acid. 4—Oleic acid. Taken from Lichtman et al. [24].

bath in which the metal was immersed. The "weakening" effect was a function of the concentration of the surface-active agent and of the chain length of the molecule. The relationship between the concentration of the solution and the change in mechanical properties is of particular interest. As shown in Fig. 11 for tin single crystals, the creep strain first increased with increasing concentration of the solution and then decreased [24]. Other mechanical parameters, such as the flow stress and stress–rupture life, follow a similar pattern.

The maximum in the curve of the change in yield strength or creep rate as related to the concentration of polar molecules was believed by these investigators [22-24] to occur when a monomolecular layer was formed. The nature of the solvent was also reported to be important.

Contrary to the results obtained by Rehbinder and co-workers, Harper and Cottrell [25] reported that the creep behavior of zinc crystals was affected by the addition of oleic acid to paraffin oil only when the specimen had an oxide surface. In Fig. 12, the creep behavior of a zinc crystal under a stress of 63.5 g/mm^2 is shown. At point A, paraffin oil was introduced, and after a 5-min immersion, the flow rate

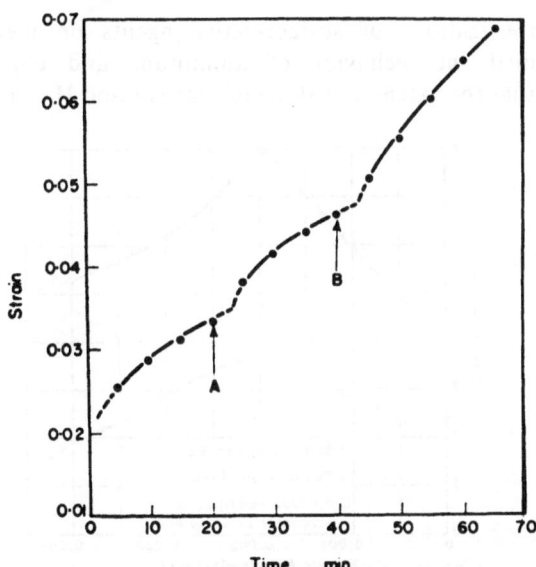

Fig. 12. Creep curves of an oxidized crystal. Paraffin was applied at point A and oleic acid solution at point B. Taken from Harper and Cottrell [25].

increased. At point B, the specimen was immersed in a 0.2% oleic acid–paraffin oil solution, and the creep rate increased further. Paraffin by itself produced an effect only in heavily oxidized specimens. With lightly oxidized specimens, the presence of oleic acid was required. Polished or etched specimens showed no response to surface-active agents. This behavior suggested to Harper and Cottrell that the effect involves the penetration of the surface film by the surface-active agent.

Klinkenberg, Lucke, and Masing [26] reported that the creep rate of gold crystals was increased when the tests were conducted in a 0.2% oleic acid–paraffin oil solution. For this investigation, the creep measurements were conducted on individual specimens; that is, a specimen was used to determine the creep curve in paraffin oil, while another specimen was tested in the solution. Since creep data are notoriously unreproducible from specimen to specimen, Kramer [27] investigated the effects of oleic acid on gold single crystals by introducing the acid into the paraffin-oil bath in incremental amounts, while the specimen was creeping at room temperature under a shear stress of 600 psi. Although the concentration of the solution during the test was changed to cover the range 0.02–4%, no change in creep rate could be detected.

In investigations of surface-active agents on metals, Kramer [27,28] studied the behavior of aluminum and copper crystals. He found that the extent and slope of Stages I and II were altered as a

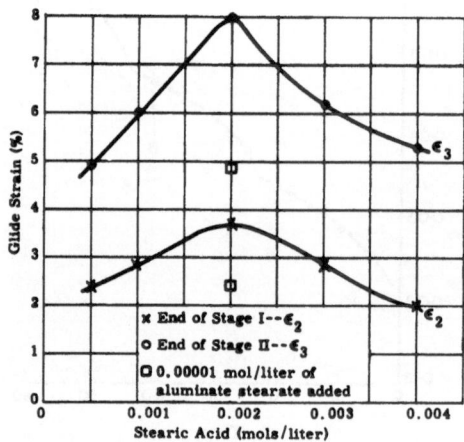

Fig. 13. Effect of concentration of stearic acid in paraffin oil on extent of Stages I and II of aluminum single crystals.

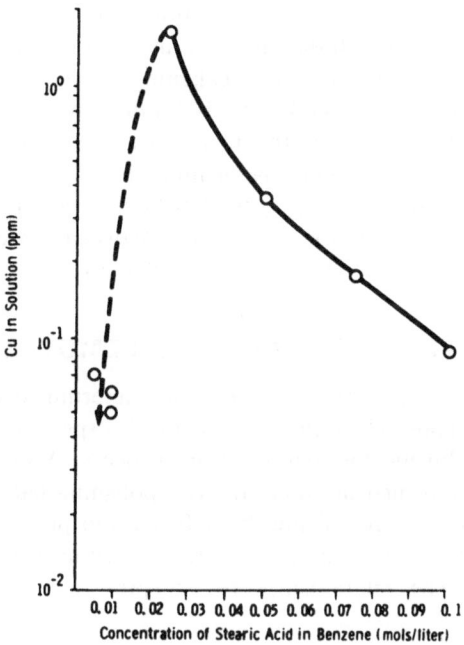

Fig. 14. The amount of copper in solution after a 3-hr immersion in stearic acid–benzene solutions of various concentrations.

function of the concentration of the surface-active agent, and the maximum weakening effect occurred for aluminum at a concentration of 0.002 moles/liter of stearic acid in paraffin oil and for copper at 0.025 moles/liter of stearic acid in benzene. Furthermore, when the solution was saturated with a metal soap of the metal tested, no change in the plastic-flow characteristics could be detected (Fig. 13). To explain this behavior and the observations that the maximum "weakening effect" is concentration-dependent, Kramer [28] determined the rate of formation of copper stearate as a function of the concentration of the benzene–stearic acid solutions in which copper single crystals were deformed. To obtain the average rate for the formation of the soap, the solutions in which the tensile specimens were pulled were analyzed after the deformation had been allowed to occur for a time of 3 hr. The resulting data (Fig. 14) showed that the maximum rate of formation occurred at the same concentration as that which produced the maximum weakening effect. Therefore, it was proposed that the change in the plastic-flow behavior of metals in solutions containing surface-active agents was associated with the rate of solution of the metal soaps which form. At low concentrations, the weakening effect is small because the rate of solution of the metal soap is limited by its rate of reaction between the surfactant and the metal. At high concentrations, the rate of formation of the metal soap is limited by the rate of the solution of the reaction products. This type of behavior would account for the extremum which exists in the relationship between concentration of the solution and the change in mechanical behavior.

EFFECTS OF SURFACE REMOVAL

In a series of investigations, Kramer determined the mechanical behavior of metals when the surface of the specimen was removed continuously during the deformation process. When metal single crystals were deformed in an electrolytic polishing bath to remove the metal at a constant rate during the deformation process, the extent of Stages I and II increased and the work-hardening coefficients decreased [29]. The relationships between the rate of metal removal

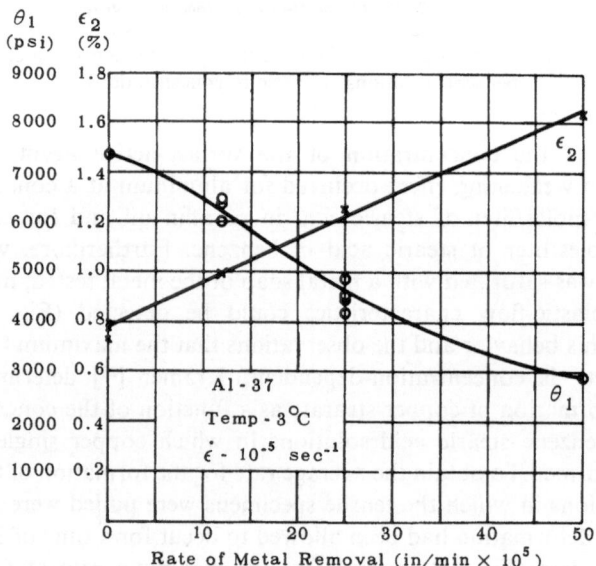

Fig. 15. The effect of rate of removal on the extent ϵ_2 and slope θ_1 of Stage I of Al-37.

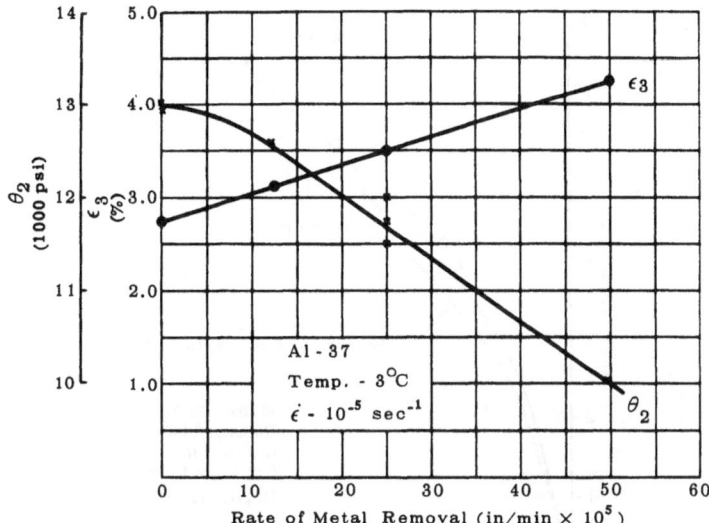

Fig. 16. The effect of rate of removal on the extent ϵ_3 and slope θ_2 of Stage II of Al-37.

and change in the extent and slopes of Stages I and II are given in Figs. 15 and 16, respectively. The plastic-flow characteristics of polycrystalline metals are also affected by a surface-removal operation during deformation. In Fig. 17, which shows the behavior of a tensile specimen of a commercial-type aluminum (1100–0), it is seen that very early in the plastic range the stress–strain curve of a specimen polished at a rate of 25×10^{-5} in./min departs from that of the specimen which was not polished. When the current through the polishing bath was reduced to zero, the work-hardening coefficient increased and the curve tended to be parallel to that of the specimen pulled at zero polishing rate. In general, the behavior of the polycrystalline specimen was the same as that of the single crystals—the work-hardening coefficient decreased with increasing rate of metal removal.

The apparent activation energy for plastic deformation was also found to be dependent upon the surface [31]. The decrease in activation energy for both single-crystal and polycrystalline aluminum is given in Fig. 18. The decrease in activation energy was found to be independent of the strain and dependent only on the rate of metal removal. At low strains, i.e., below about 4% for the orientation of the crystal used in the investigation, the activation energy was 4200 cal/mole. At a

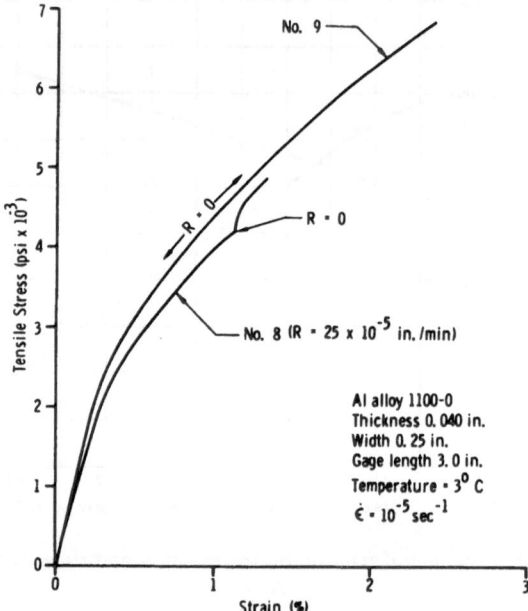

Fig. 17. Stress–strain curve for a commercially pure aluminum (1100-0) deformed
while the surface was removed at a rate of 25×10^{-5} in./min.

polishing rate of 50×10^{-5} in./min, the activation energy was reduced
to 920 cal/mole. The change of the activation energy for high-purity
polycrystalline aluminum as a function of polishing rate appears to be
the same as that for single crystals. The change in activation energy
with rate of metal removal for gold, aluminum, and copper is given
in Fig. 19. It is seen that the effect of surface removal is largest for
aluminum and smallest for gold.

Not only is the activation energy affected by the surface, but the
activated volume is increased when the surface is removed during the
deformation process. Figure 20 shows the effect of polishing on the
activated volume V, where

$$\beta = \frac{V}{kT} = \frac{\Delta \ln \dot{\gamma}}{\Delta \sigma_a}$$

The measurements were obtained by changing the strain rate $\dot{\gamma}$ by a
factor of ten and noting the increase in stress $\Delta \sigma_a$ which occurred when

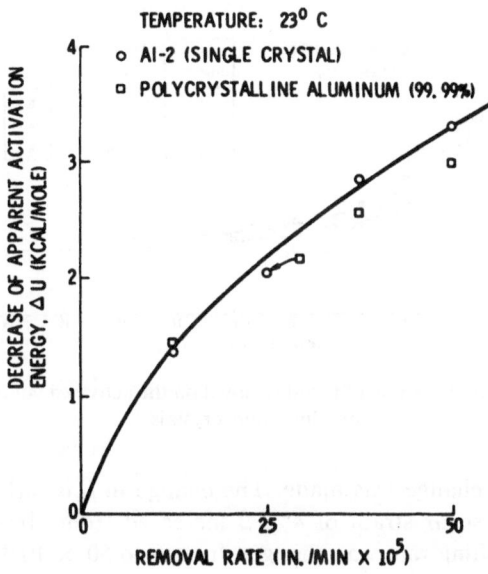

Fig. 18. The change in apparent activation energy ΔU as a function of the rate of metal removal for single-crystal and polycrystalline aluminum.

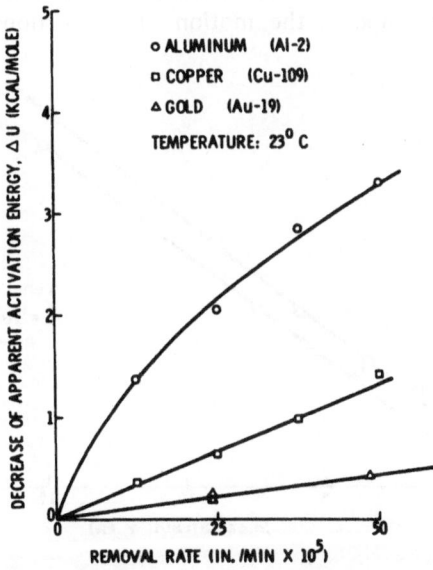

Fig. 19. The change in the apparent activation energy ΔU of single crystals of aluminum, copper, and gold as functions of rate of removal R.

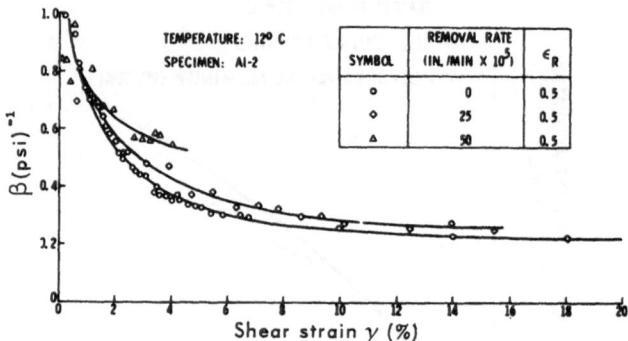

Fig. 20. The effect of the rate of metal removal on the activated volume ($V = \beta kT$) for aluminum crystals.

the strain-rate change was made. The change in β is rather large; for example, at a shear strain of 4%, β increased from about 0.4 to 0.6 when the polishing rate was changed from 0 to 50×10^{-5} in./min.

To explain the various surface effects, Kramer [30] proposed that the dislocation density in regions near the surface of a deformed specimen is greater than that in the interior. This layer, referred to as the debris layer, impedes the motion of dislocations. Accordingly,

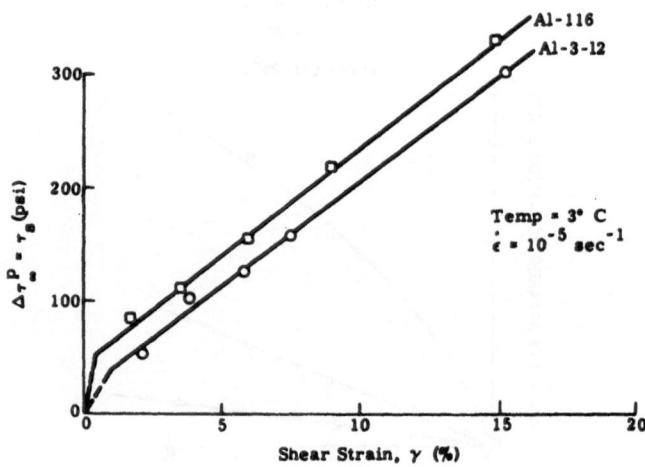

Fig. 21. Relationship between τ_s and shear strain γ for specimens Al-3-12 and Al-116.

the net stress τ^* acting on a dislocation may be expressed as follows:

$$\tau^* = \tau_p - \tau_i - \tau_s$$

where τ_p is the applied plastic shear stress, τ_i is the stress field due to internal obstacles which must be overcome by the moving dislocation, and τ_s is the stress field which must be overcome when the dislocations pass through the debris layer. The value of τ_s was measured by deforming specimens to a given strain and noting the decrease in the initial flow stress after the load was reduced to zero and the debris layer removed by electrolytic polishing. The results of Figs. 21 and 22 show that τ_s or σ_s (normal stress) are linear with respect to the strain.

It appears possible to explain many of the environmental effects of metals in terms of the surface stress τ_s. This surface stress should be a function of the accumulation of dislocations in the surface layer. Therefore, any coating, such as an oxide or metal film, which would impede the egress of dislocations would also increase τ_s. In a vacuum or an inert atmosphere, the rate of formation of an oxide film on the freshly

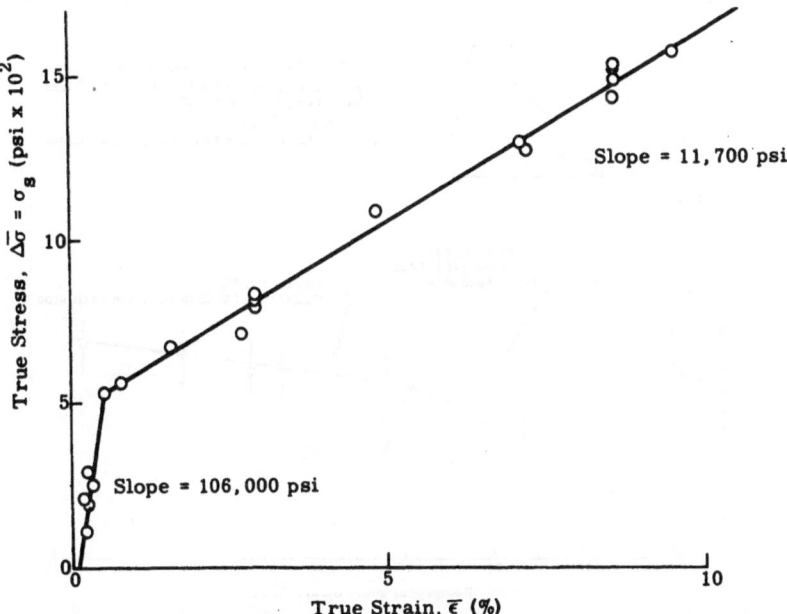

Fig. 22. Relationship between σ_s and strain in high-purity polycrystalline aluminum.

exposed metal at the slip steps would be slow, and the rate of escape of dislocations would be correspondingly higher. This would account for the higher creep rate of specimens in vacuum and the lower creep rate of coated specimens. It is not to be supposed that the films are solely responsible for the accumulation of dislocations in the surface layers of a deformed specimen. A debris layer has been found in gold—a metal which does not have an oxide film. The mechanism of formation of the debris layer in metals without an oxide film is not clear, but it may be the result of the formation of tangles or loops when the dislocations emerge. In any case, there are at least two factors which promote the formation of the debris layer: (1) the surface itself, and (2) the presence of a strong film acting in conjunction with the surface. Therefore, it would be expected that the effect of a vacuum environment, for example, on the creep behavior would decrease with strain. Such an effect is shown in Fig. 23 for aluminum single crystals. In this graph, the change in activation energy was determined by measuring the change in creep rate when the pressure was changed quickly between 10^{-7} and 2×10^{-4} torr. Initially, the activation energy changed by about 500 cal/mole; however, with continued deformation, the change

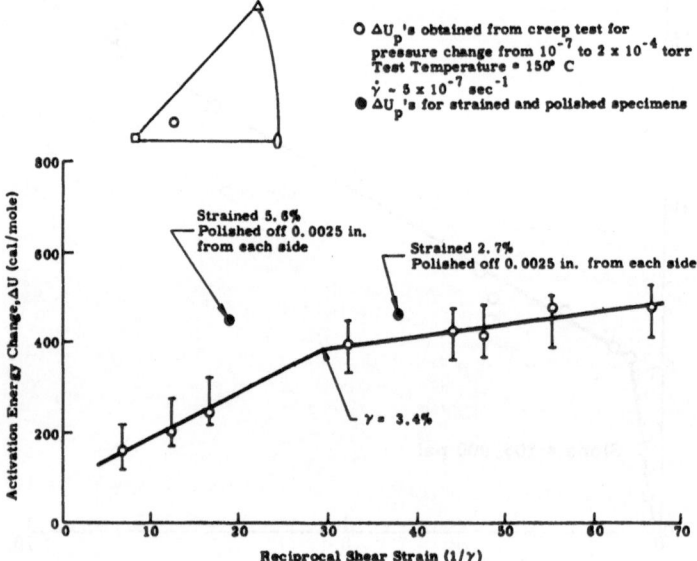

Fig. 23. Change in activation energy for aluminum single crystal as a function of strain.

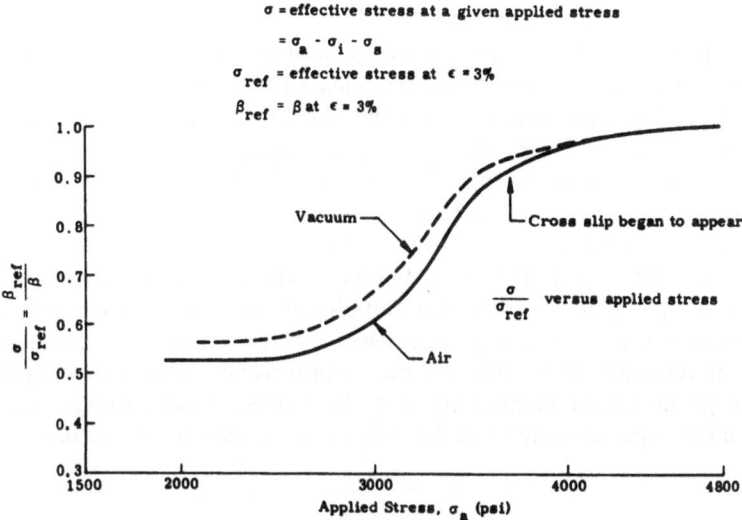

Fig. 24. Effective stress ratio as a function of applied stress in air and in vacuum for high-purity polycrystalline aluminum.

became very small. At strains of 2.7 and 5.6%, the specimens were polished to remove the debris layer, and it is noted that the change in activation energy was the same as that of the virgin specimen. From this concept of the debris layer, it would also be expected that an inert atmosphere would be effective only in changing the mechanical behavior of metals whenever the oxide has a large influence on the formation of the surface layer and, therefore, on τ_s. For polycrystalline metals, as shown in Fig. 22, σ_s increases very rapidly with strain, and, therefore, for high strains, any atmospheric effect may be expected to be small. This effect has been observed experimentally in tensile tests on aluminum and iron specimens [32]. Practically no change in mechanical behavior occurred in the range of pressures from 760 to 10^{-8} torr. The sensitivity of the strain measurement was too low to detect changes at low strains. However, in creep tests when the strain rate was of the order of 10^{-7} sec^{-1}, the changes could be detected. These data are shown in Fig. 24 for high-purity aluminum in terms of the activated volume ($V = \beta kT$). At low stresses (or strains), there is a small but definite separation between the values of β obtained from measurements made in air and at 10^{-6} torr; at the high stresses, the curves are coincident.

CONCLUSION

It can be stated that the creep, fatigue, stress–rupture, and the tensile behavior of metals are influenced by the conditions that exist at the surface and in the region within the specimen close to the surface. The surface effects are important from both practical and theoretical viewpoints. Failure to take surface effects into account and complete reliance upon explanation based on internal dislocation obstacles lead to inconsistencies. It is obvious that vacuum or inert-gas effects can not be interpreted in terms of internal dislocation mechanisms. Of importance from a practical standpoint are the observations that the fatigue life is increased greatly when reactive species are eliminated from the atmosphere, whereas the creep resistance and stress–rupture life are decreased. Conditions that allow dislocations to escape easily from the specimens tend to decrease the resistance to plastic flow.

REFERENCES

1. R. Roscoe, *Nature* **133**: 912 (1934).
2. A. H. Cottrell and D. F. Gibbons, *Nature* **162**: 488 (1948).
3. S. Harper and A. H. Cottrell, *Proc. Phys. Soc. (London)* **B63**: 331 (1950).
4. J. Takamura, *Mem. Fac. Eng. Kyoto Univ.* **18**(3): 255 (1956).
5. M. R. Pinkus and E. R. Parker, *Trans. AIME* **191**: 792 (1951).
6. H. J. Gough and D. G. Sopwith, *J. Inst. Metals* **49**: 93 (1932).
7. H. J. Gough and D. G. Sopwith, *J. Inst. Metals* **56**: 55 (1935).
8. H. J. Gough and D. G. Sopwith, *J. Inst. Metals* **72**: 415 (1946).
9. N. J. Wadsworth, *Proceedings of the Symposium on Internal Stresses and Fatigue in Metals*, Elsevier Publishing Co. (Amsterdam), 1959, p. 382.
10. K. V. Snowden and J. N. Greenwood, *Trans. AIME* **212**(5): 626 (1958).
11. K. V. Snowden, *Nature* **189**: 53 (1961).
12. K. V. Snowden, *Acta Met.* **12**: 295 (1964).
13. S. E. Podlaseck, H. Shen, and I. R. Kramer, unpublished results, 1963.
14. M. R. Achter, G. J. Danek, and H. H. Smith, *Trans. AIME* **227**: 1296 (1953).
15. N. Thompson, N. Wadsworth, and N. Louat, *Phil. Mag.* **1**: 113 (1956).
16. E. R. Mantel, G. H. Robinson, and R. F. Thompson, *ASM* **1**: 57 (1961).
17. H. E. McCoy, Jr., and D. A. Douglas, Jr., ORNL Rept. 2972, September 5, 1962.
18. P. Shahanian, *Trans. Am. Soc. Metals* **49**: 862 (1957).
19. P. Shahanian and M. R. Achter, *Trans. Am. Soc. Metals* **51**: 244 (1957).
20. P. Shahanian and M. R. Achter, *Proceedings of Joint International Conference on Creep*, 1963.
21. I. R. Kramer and S. E. Podlaseck, *Acta Met.* **11**: 70 (1963).
22. P. A. Rehbinder, *Byul. Akad. Nauk Classe. Sci. Mat. Nat. Ser. Chem.* 639–704 (1936).

23. P. A. Rehbinder and E. K. Wenstrom, *Byul. Akad. Nauk URSS Classe. Sci. Mat. Nat. Ser. Phys.* 531–548 (1937).
24. V. I. Lichtman, P. A. Rehbinder, and G. V. Karpenko, "Effect of Surface-Active Medium on the Deformation of Metals," Her Majesty's Stationery Office (London), 1958.
25. S. Harper and A. H. Cottrell, *Proc. Phys. Soc. (London)* **B63**: 331 (1950).
26. V. W. Klinkenberg, K. Lucke, and G. Masing, *Z. Metallk.* **44**: 362 (1953).
27. I. R. Kramer, *Trans. AIME* **227**: 529 (1963).
28. I. R. Kramer, *Trans. AIME* **221**: 989 (1961).
29. I. R. Kramer and L. J. Demer, *Trans. AIME* **221**: 780 (1961).
30. I. R. Kramer, *Trans. AIME* **222**: 1003, 1963 (1963).
31. I. R. Kramer, *Trans. AIME* **230**: 991 (1964).
32. I. R. Kramer, H. Shen, and S. E. Podlaseck, AFOSR Rept. 64–2509.

Ultrafine Particles in the Gas Phase

John Turkevich

Chemistry Department
Princeton University
Princeton, New Jersey

INTRODUCTION

During the last decade, electron microscopy was used to study the various catalytic material preparations as well as related colloidal preparations. Quantitative studies of the various factors that determine the size, shape, and uniformity of ultrafine particles in solution made possible a chemical formulation of the process of particle formation in solution and a reproducible procedure for catalyst preparations. A very important concept came out of this work; namely, the particle-size distribution curve as determined by electron microscopy is *not* an error curve, but it does contain information about the processes of nucleation, growth, and aggregation of ultrafine particles.

The success achieved in dealing with ultrafine particles in the liquid phase encouraged investigation of ultrafine particles in the gas phase. We were primarily interested in inorganic solids and not in liquids, such as fogs or mists. Our interest from a practical viewpoint was to determine by electron microscopy the nature of air contamination; nevertheless, from an academic viewpoint, we were and still are interested in how a disordered array of gaseous atoms condenses to form an ordered array of ultrafine particles.

EXPERIMENTAL

It is now possible to speculate on the definition of an ultrafine particle. Is there a continuous gradation of sizes from particles composed of one atom to particles composed of ten thousand atoms? Both theoretical and experimental evidence seems to indicate that there is a lower limit of size below which a particle does

195

not exist, as evidenced by the fact that the surface tension of the particle tends to disrupt the particle, while the lattice energy tends to hold the particle together. Since the surface energy varies directly with the square of the radius, while the lattice energy varies directly with the cube of the radius, there is a discontinuity at the radius where these two are equal. At lower radii, the particles are unstable, whereas at higher radii they are stable and can grow. The critical radius where these two energies are equal determines the size of the nucleus of the ultrafine particle. Thus, in an ultrafine particle, one is dealing with an entity more complex than an atom or a molecule—an entity which is endowed with a new set of properties, those associated with the texture of the material. Texture involves the size, shape, and state of aggregation of primary ultrafine particles, and it also involves defects and dislocations in the ordered array of component atoms or molecules in the particle. Furthermore, texture endows the particle with interesting surface, mechanical, and electrical properties. Accordingly, control of the texture of substances offers a new challenge for the scientist and technologist dealing with materials.

Over the course of a century, scientists have learned the chemistry of the elements and the behavior of atoms and have obtained detailed information about the reactions of compounds and the structure of molecules. As a result of these achievements, a new era of materials research and development has been initiated—an era that is concerned with the synthesis of materials with known and predetermined texture. In the future, work will be performed with an organization of matter more complex than that found in molecules.

It is generally assumed that this next stage of organization of matter is disordered and follows laws of statistics, rather than obeying well-controlled processes. It is also assumed that, with the exception of the process of crystallization, the piling up of atom on atom and molecule on molecule is a haphazard process. However, there is evidence that natural laws exist which govern the synthesis of texture; under these laws, the assumption of haphazard pileup of atom on atom or molecule on molecule would be proven false. There must be laws of synthesis of size, shape, particle-size distribution, incorporation of impurities, and defects just as there are laws of synthesis of molecules. It is our task to uncover these laws and apply them to the synthesis of materials.

An important aspect of the texture of a material is its capability of storing energy and information. Catalysts are substances that control chemical reaction, indicating to the reactants which of the many thermodynamically possible paths the reactants should take. This catalytic action is effected by specific adsorbability on surfaces (a process analogous to a lock and key) and by specific energy transfers. A nucleus is a catalyst for its own growth; the more chemically complicated the ultrafine particle, the more complicated the nucleus. Thus, both catalysts and nuclei of particles must be complicated structures, involving a much higher degree of complication than is found in atoms of hydrogen, helium, or in molecules such as methane and carbon tetrachloride. The ultimate in useful complication is the DNA molecule, which is capable of storing information that leads to the synthesis of a biological individual.

The complexity of and the possibility of detail in the texture of material provides the possibility of storing information in material without affecting its gross chemical properties. Furthermore, the very existence of texture indicates that, somewhere in its production, detailed information was imparted to the material. Thus, an ultrafine particle of iron oxide (Fe_3O_4) smoke is a prism that is flat on its two bases and has six sides. In order to produce this morphology and texture, a nucleus had to be formed not of two or three atoms, but of dozens of atoms, because the unit cell of this material is large. This process has yet to be elucidated, but certainly some ordering process took place in order to produce a nucleus of a finely divided aerosol to make a catalyst for its own growth. Thus, production of catalysts for chemical reaction is related to the production of nuclei for the growth of ultrafine particles.

The various methods of preparation of ultrafine particles in the gas phase can be divided into two classes—dispersion and aggregation. The dispersion methods use localized energy to first break up the material into fine particles and then to stabilize the particles by giving them an electric charge. Passage of a pulse of current through a fine wire seemed an elegant method of producing ultrafine particles in the gas phase, but, upon electron microscopic examination of the particulate matter produced, it was seen that the aerosol consisted of irregularly shaped boulders which were thrown off from the exploding wire. Another method of production of ultrafine particles involves breaking up a stream of material coming from a fine nozzle by imparting to the stream a high electric potential. This method of dispersion, however, is limited to materials that can be obtained readily in a molten state. In general, the disintegration methods are of minor interest to us, since they involve the disruption of material that has a complicated and unknown structure. The aggregation methods, on the other hand, are of greater interest. The white deposit found on reagent bottles of most chemical laboratories is the result of an aggregation process in which ammonia reacts with hydrogen chloride to form ammonium chloride molecules, which aggregate to a smoke and settle all over the laboratory.

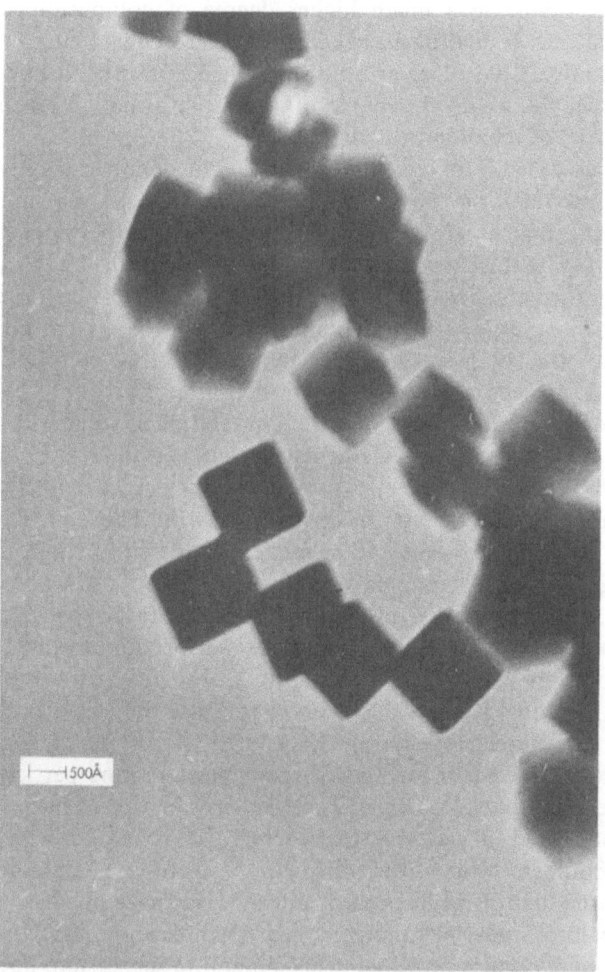

Fig. 1. Magnesium oxide smoke. Aerosol from DC arc. Magnesium electrodes. 200,000 × ; reduced 50 % for reproduction.

Another method of producing finely divided particles by aggregation is the one developed by Langmuir, La Mer, and Sinclair for producing smokes in chemical warfare. Also, ultrafine particles can be produced by aggregation of products of combustion and by photolysis of materials such as iron carbonyl. The two aggregation processes for production of ultrafine particles in the gas phase that we have investigated in detail are vaporization and aggregation of particles either in a DC electric arc in air or in an electric heater in an atmosphere of helium.

Those who have had experience in emission spectroscopy will recall that a DC arc produces not only an intense light characteristic of the electrodes, but also a smoke characteristic both in shape and chemical composition of the particulate matter. Collecting smoke from electrodes of various elements at a fixed distance from the arc on a collodion membrane of the type used in electron microscopy, Amick and Turkevich found that the particulate matter was characterized for chemical composition by electron diffraction and for shape and particle-size distribution by electron microscopy. We concluded, from a survey of thirty-two elements, that the shape of the ultrafine particles of the smoke produced in the DC arc was often characteristic of the chemical nature of the electrode and that this shape became all the more distinctive as the particle size became small and approached the size of a small number of unit cells. Thus, the morphology and the texture may be used for analytical identification by measuring the ratio of the lengths of the three sides and the angles between these sides.

Magnesium electrodes produce cube-like particles of magnesium oxide characterized by a thread-like aggregation in which the edge of one cube is attached to the edge of another cube (Fig. 1). The zinc electrodes give a zinc oxide smoke the particles of which consist of spikes arranged in tetrahedral array around a central crystallite. This shape is characteristic of zinc oxide (Fig. 2). Further data are listed in Table I.

This survey of the shape of ultrafine particles produced by vaporization of material in the gas phase suggests that they have characteristic shapes determined by the chemical composition of the particle. These shapes will be the more distinct and characteristic as the particle size becomes smaller and approaches that of the unit cell. Electron microscopy permits, by shadowing techniques and stereoscopic measurements, the determination of the axis ratios and the axis angles that are characteristic of the material. Thus, electron microscopy offers a very sensitive method of determining the composition of

Fig. 2. Zinc oxide smoke. Aerosol from DC arc. Zinc electrodes. 400,000 × ;
reduced 50% for reproduction.

individual ultrafine particles. A further refinement of the identification
procedure would be to effect known chemical transformations of the
individual particles and observe changes in morphology. Gaseous
reagents, such as oxygen, hydrogen, chlorine, acetylene, and butene,
could be used. Enhanced reactivity could be obtained at room temperature by using hydrogen atoms, oxygen atoms, *etc.*, produced in an

Table I

Data on Particles of Smoke Produced in a DC Arc

Compound	Crystallographic modification	Size	Remarks
Chromium	12-Sided plates	Diameter, 500–1500 Å; thickness, 75 Å	Electron diffraction pattern identifies the ultrafine particles as Cr_2O_3
Indium	Rhombohedral	Length, 100–1000 Å	Of In_2O_3
Manganese	Cubes with puffed-out sides	Particle edges, 100–1000 Å	Of Mn_3O_4
Iron	Hexagonal drums	Diameter, 25–500 Å	Of Fe_3O_4
Molybdenum	Hexagonal plates	Diameter, 500–2000 Å; thickness, 75 Å	
Nickel	Cubes, usually congregated into threads	Diameter, 50–1000 Å	Of NiO
Lead	Elliptical plates	Length, 75–10,000 Å; thickness, 100 Å	Of PbO
Tin	Rhomboid and cubics in its projections of the supporting collodion membrane; particles linked together into long threads	25–1000 Å	Of SnO_2
Titanium	Hexagonal particles	Diameter, 25–500 Å	Of TiO_2 (anatase)
Thorium	Cubic particles with rounded corners, with tendency to form threads	25–500 Å	Of ThO_2
Uranium electrodes	Finely divided smoke; individual particles giving hexagonal and square projections		Of UO_3
Tungsten	Rhomboidal	Uneven particle-size distribution of a few particles with diameters below 100 Å and most in the range of 200–500 Å	Of WO_3

electric discharge. A development of such a method may be the ultimate in sensitivity for a chemical analysis.

One of the troublesome problems in aerosol work is the aggregation of the primary ultrafine particles. Growth is slow in the gas phase because the nuclei tend to be perfect with few dislocations and imperfections. In solutions, these develop on the nuclei because of the surface adsorption by the nuclei of the numerous impurities always present in solution. The impurities are usually present in much smaller

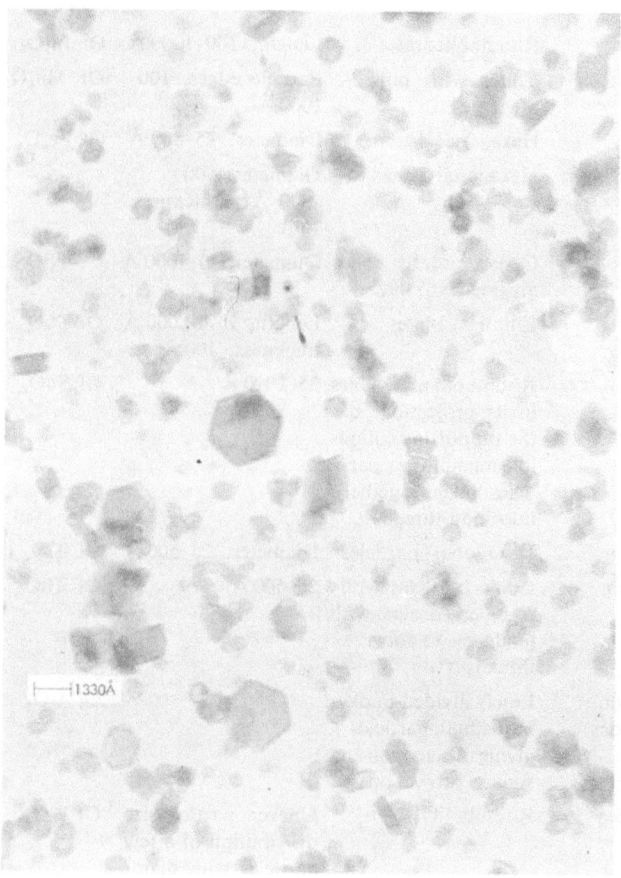

Fig. 3. Magnesium crystals grown in helium formed by evaporation at 1400°C in 1 mm of helium and collected 9.2 cm from source. Serial No. 3-3-55-2-C. 75,000 × ; reduced 50% for reproduction.

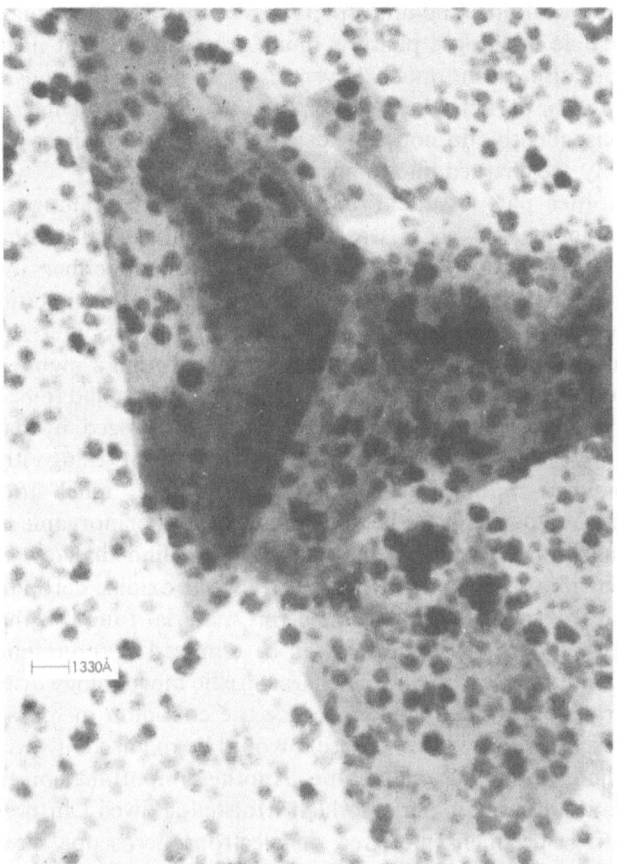

Fig. 4. Magnesium crystals grown in the presence of copper. Serial No. 6-24-55-1-C.
75,000 × ; reduced 50% for reproduction.

concentrations in the gas phase, however. In fact, in most gas-phase synthesis of crystals, the growth process takes place on the walls of the container.

Till and Turkevich were able to show the effect of impurities on the growth of magnesium crystals. When magnesium is evaporated in helium, very small crystals are formed on condensation of the vapor (Fig. 3). On the other hand, when it is evaporated in helium in the presence of heated copper, large plates of magnesium are formed (Fig. 4). Thus, the growth process is slow in the gas phase. On the other hand, aggregation is favored for two reasons. The Brownian motion of a particle in the gas phase is much greater than that in the liquid phase because of the much lower viscosity of the gaseous medium. Furthermore, particles in the gas phase usually do not have a charge and, consequently, do not repel each other, while particles in solution invariably are charged due to adsorbed ions.

Another part of the Princeton research program on ultrafine particles in the gas phase was the examination of ultrafine particles present in Princeton air. Collections were made either by natural settling of particles on collecting pans, by thermal precipitation, or by filtration using millipore filters. Although primarily interested in the morphology of the inorganic constituents, we found a high proportion of carbonaceous deposits present in urban air, which had to be removed before the inorganic material could be characterized by its texture. Removal of this carbonaceous matter by oxidation (using either air or pure oxygen) requires a temperature of 500°C, much too high a temperature to preserve the morphology of the inorganic materials which may either volatilize or sinter. We did find that oxygen atoms produced in a Wood's electric discharge will oxidize carbon at room temperature. In this way, carbonaceous material found in the aerosol or fiber material from the filters can be removed at room temperature with minimum distortion of the characteristic morphology or texture of the inorganic ultrafine particles. Since the collodion membrane could not be used for this work, because it would be oxidized, it was replaced by a silicon monoxide membrane. Another important constituent of aerosols is siliceous material, which Molsted showed can be removed by treating specimens mounted on electron-microscope screens with hydrogen fluoride gas at 100°C. Further characterization of the material can be obtained by reduction of transition-metal oxides or fluorides with hydrogen atoms, by volatilization of nickel with carbon monoxide, and by the detection of copper particles by their capacity to produce threads of cuprene with acetylene gas. Reagents used for this work must be gaseous, since liquid reagents would tend to agglomerate the ultrafine particles mounted on the membrane when the liquid is evaporated prior to examination in the electron microscope.

In a quantitative study of the kinetics of the formation of metallic smoke made at Princeton by Till and Turkevich, a known amount of metal was placed in a tungsten coil in a large bell-shaped vacuum evaporator (Fig. 5). Collodion screens, such as used for mounting specimens for electron-microscope examination, were placed at various distances

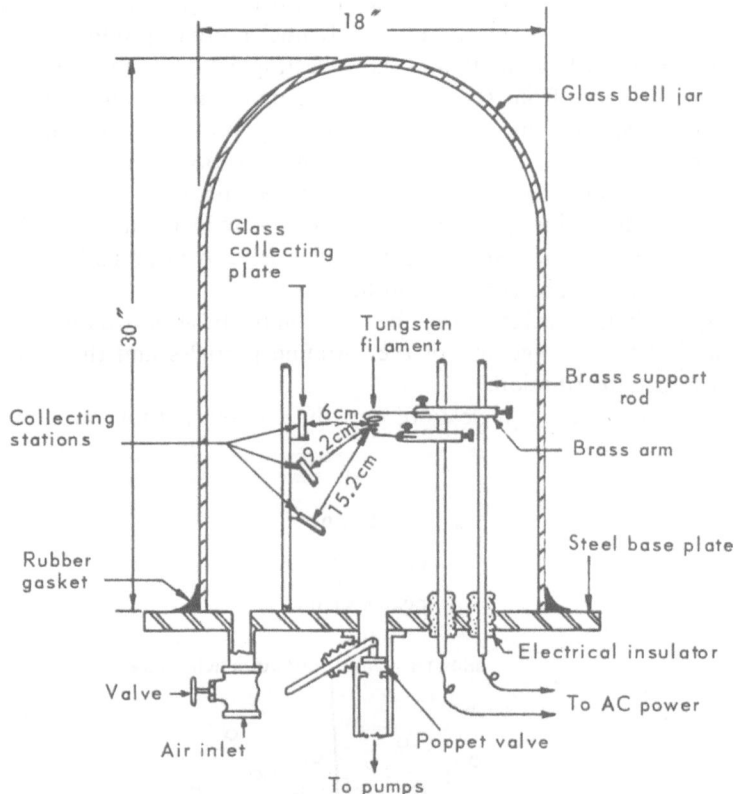

Fig. 5. Apparatus for evaporation of metals in a controlled atmosphere.

from the tungsten coil. The metal was evaporated by passing an electric current through the tungsten coil in a high vacuum of 10^{-6} mm Hg; this caused the evaporated metal to travel in a straight line and condense on the collodion collecting screen to form an apparently uniform deposit at a resolution of 15–20 Å of the electron microscope used at that time. Whatever structure and particulate matter formed on the surface was due to surface migration on the collodion screen. A check for this surface-migration effect, when it did occur, was the identical appearance of the deposit at various distances from the source. Nucleation growth and aggregation of the particles was induced in the gas phase by introducing an increasing amount of helium gas into the bell jar. As the pressure of helium increased, the evaporated atoms did not travel in a straight line, but, due to collision with the helium gas, traveled in jogs, colliding with each other and ultimately nucleating to produce a finely divided particle. Examination of the collecting collodion membranes at various distances from the source of the metal atoms revealed three concentric spheres in which the following processes predominate: atomic evaporation, nucleation and atomic evaporation, and growth and aggregation. The three spheres can be increased in radius by decreasing the helium pressure and decreased by increasing the helium pressure (Fig. 6). These spheres can be identified by examination of the electron micrographs (Figs. 7 and 8).

Quantitative evaluation can be made on the basis of two observable quantities: the average size of the ultrafine particles and the detailed particle-size distribution.

The processes of nucleation, growth, and aggregation can now be

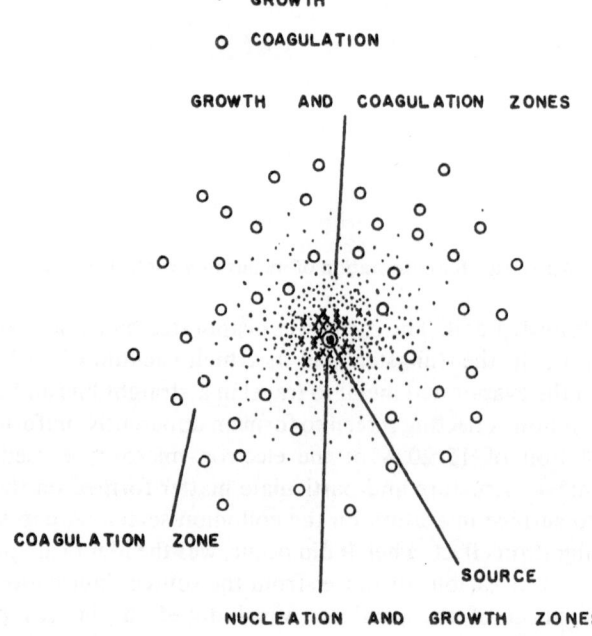

Fig. 6. Zones of ultrafine-particle formation around heated source of metal atoms.

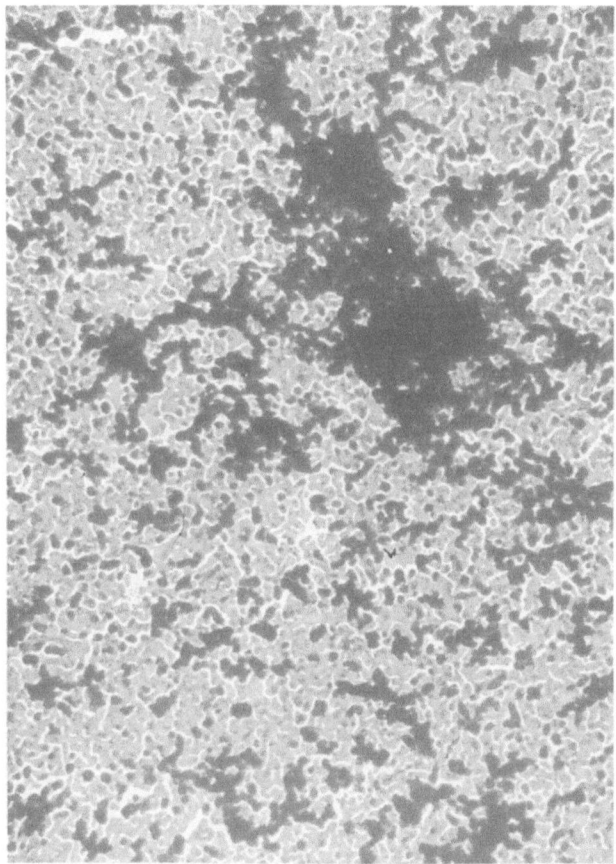

Fig. 7. Silver smoke formed at 1.0 mm helium pressure and collected 9.2 cm from source, indicating atomic deposition, collecting membrane, and nucleation and aggregation in the gas phase.

considered from the quantitative point of view, with the free energy of formation of a spherical nucleus of radius r from the vapor written as the sum of a favorable bulk term and an unfavorable surface term. Thus,

$$\Delta F = -\frac{4\pi r^3}{3}\frac{\rho}{m} RT \ln S + 4\pi r^2\sigma \tag{1}$$

when ρ is the density, m the molecular weight, σ the surface free energy per unit area, and S the degree of supersaturation defined as the ratio of the vapor pressure at the experimental conditions to the equilibrium pressure at the same temperature over a large surface. The plot of ΔF as a function of the radius for values of S greater than unity increases to a maximum positive free energy and then rapidly decreases to favorable negative free energies. The critical nuclear radius r_c is found by maximizing ΔF with respect to r and can be expressed as follows:

$$r_c = \frac{2\sigma m}{\rho RT \ln S} \tag{2}$$

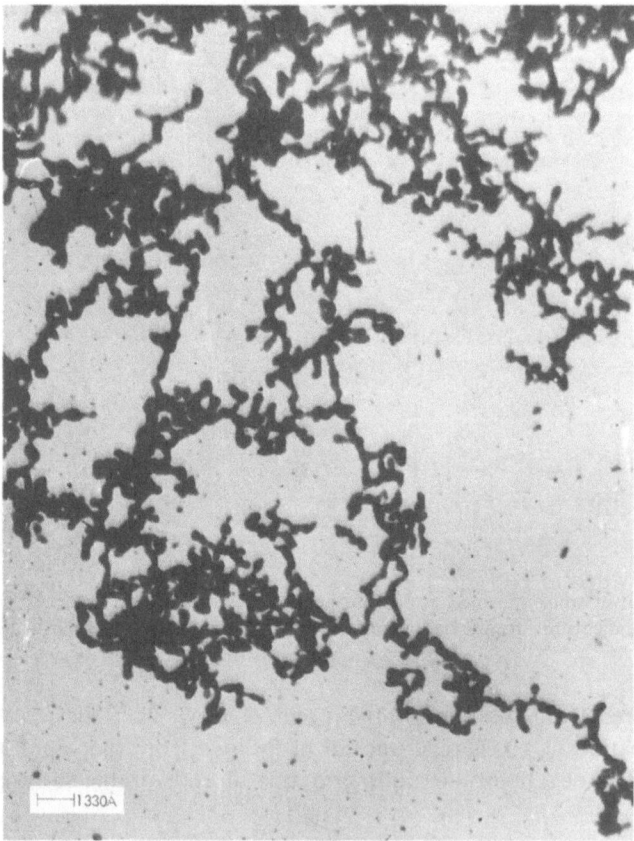

Fig. 8. Silver smoke formed at 11.0 mm helium pressure and collected 6.0 cm from filament source, indicating nucleation and aggregation in the gas phase. 75,000 × ; reduced 50% for reproduction.

and

$$\Delta F = \frac{16m^2\sigma^3}{3\rho^2(RT \ln S)^2} \tag{3}$$

If it is assumed that the temperature of the nucleation zone is 400°K and the average partial pressure of the metal in the vapor is 0.1 mm Hg, then the degree of supersaturation is determined by the equilibrium value of the vapor pressure of the element at 400°K. The surface energy of the nucleus is taken as that of the liquid metal. Calculation shows that, for gold, copper, silver, manganese, and antimony, the number of atoms in the nucleus is one; while, for magnesium, zinc, and cadmium, it is three. This small value may be due to an incorrect value for the surface energy of a small particle.

Becker and Doring have developed an expression for the rate of formation of stable nuclei using a combination of thermodynamic and kinetic methods which in a simplified form is given by the following equation:

$$I = Ze^{-\Delta F^*/kT} \tag{4}$$

where Z is the number of binary collisions per unit volume per second of the metal atoms and ΔF^* is the height of the free energy maximum in nucleus formation, a quantity which is highly dependent on the degree of supersaturation. Since the evaporation was adjusted so as to be roughly the same for different metals, the metal vapor pressure was approximately constant in the different evaporations. The variation in the average value of S for different metals should be determined primarily by the equilibrium vapor pressures of the metals at the average nucleation temperature. Thus, the ΔF^* and the rate of nucleation are determined by this equilibrium vapor pressure. It can be shown that the final diameter of the particle is inversely proportional to the rate of nucleation. The faster the rate, the smaller the diameter of the ultimate particle size produced. We obtain the following relation:

$$\bar{D}^3 = Ae^{\Delta F^*/kT} \tag{5}$$

where $\bar{D}$ is the average diameter of an aerosol. Figure 9 is a plot of log $\bar{D}$ *versus* $\Delta F^*/kT$, where it is assumed that the surface tension of the metal is that of the solid. Two linear plots are obtained differing in slope and displaced from each other. Failure to obtain a single line may be due to a change in volume of the nucleation zone, which was assumed constant for all classes of metal, or to a variation of the surface energy with particle size.

A deeper insight into the nucleation process can be obtained from the analysis of the particle-size distribution curves. Most of the particle-size distributions observed (notably for silver, gold, and antimony) are unsymmetrical, with a skewed tail in the direction of the larger sizes (Fig. 10). However, the size distribution of the aluminum aerosol is symmetrical.

We shall now consider how we can determine a nucleation curve from a size-distribution curve. The growth process takes place by collision of metal atoms in the gas phase with the nucleus and can be shown to be governed by the following law:

$$\frac{dD_i(t)}{dt} = kn(t) \tag{6}$$

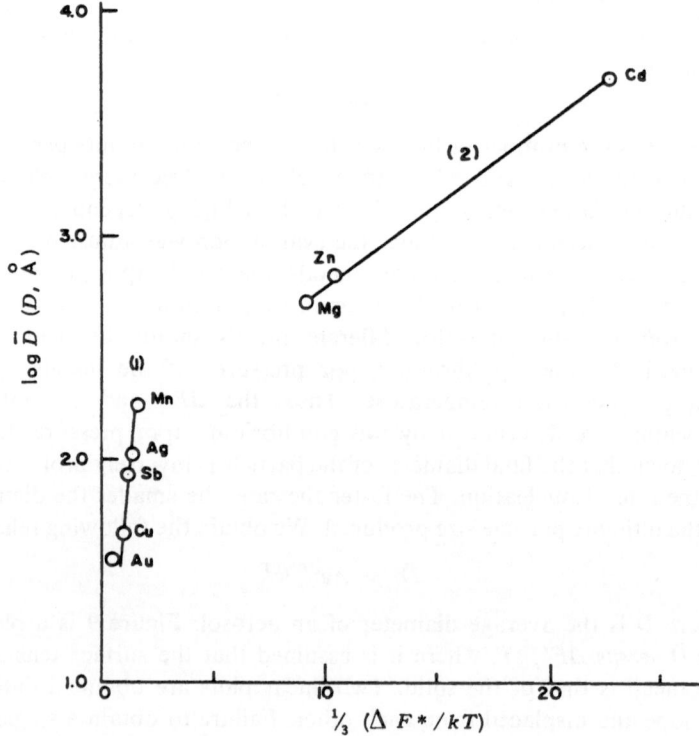

Fig. 9. Relationship between the average diameter $\bar{D}$ (in angstroms) and the free energy of nucleation.

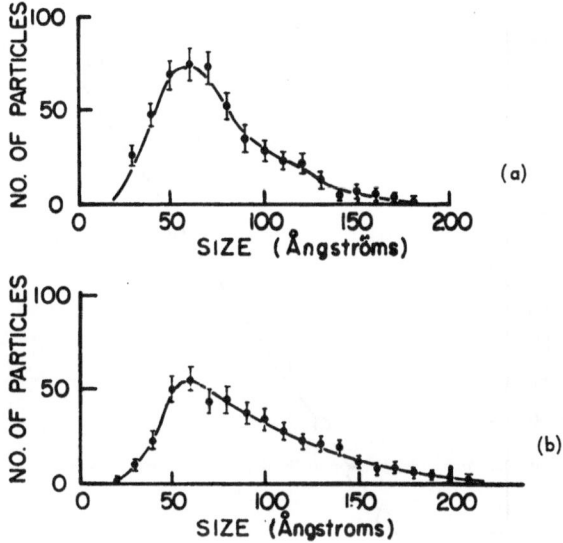

Fig. 10. Particle-size distribution of silver at 5.0 mm (Hg) He. (a) Collected 6.0 cm from filament. (b) Collected 9.2 cm from filament.

where $D_i(t)$ is the diameter of the ith particles at time t and $n(t)$ is the concentration of metal ions around the particle i at time t. Integration of equation (6) gives the following expression for obtaining a nucleation curve from the particle-size distribution:

$$t_n = \frac{D_F - D_c}{kn_0} \ln \frac{D_F - D_c}{D_{it} - D_c} \tag{7}$$

where t_n is nucleation of the ith particle; k is a constant; n_0 is the metal concentration of gaseous metal atoms; D_F is the diameter of the largest particle; D_c is the diameter of the nucleus; and D_{it} is the diameter of the particular size in the size distribution. The nucleation curves were obtained by calculating the value of kn_0t_n for the size of particle i and by plotting this value against the cumulative percentage of the population of the size distribution up to and including particle i, starting with the largest particle in the distribution.

The nucleation curves of silver, gold, and antimony plotted on probability graph paper are linear plots, indicating normal distribution (Fig. 11). It is seen that the unsymmetrical size-distribution curve

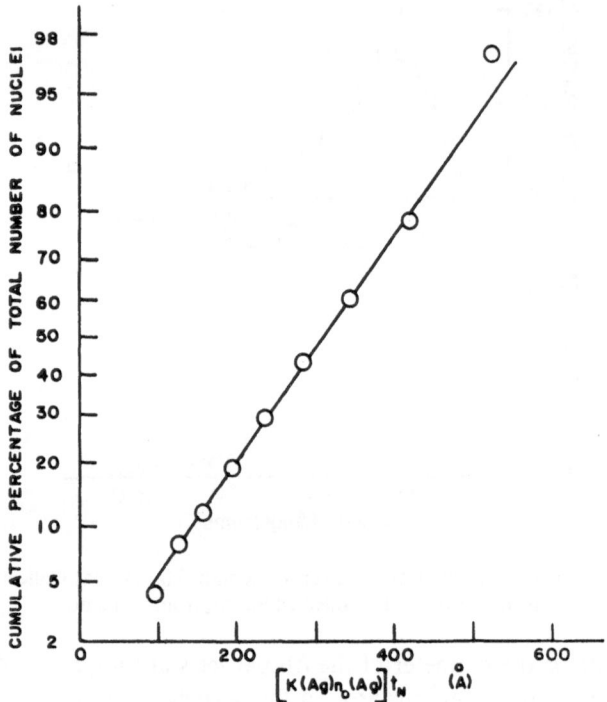

Fig. 11. Nucleation-rate curve for silver.

results from a symmetrical, normally distributed nucleation process. On the other hand, similar analysis of the symmetrical size-distribution curves of aluminum and magnesium indicates that there are two nucleation processes involved. These may be associated with two types of nuclei—a metal nucleus and an oxide-type nucleus.

CONCLUSION

Materials research is moving more and more into the field of synthesis of materials with predetermined structure—particle size and shape; aggregation of primary building blocks; and nature of defects, both chemical and physical. Electron microscopy offers an important tool in elucidation of the mechanisms involved in these syntheses.

Panel Discussion: Session II

N. W. Huberman (Atomics International): With regard to impurities at a partial pressure of 10^{-8} mm, a monolayer of oxygen can be adsorbed in 100 sec. Would not a test of the role of impurities in surface adsorption be to go down in pressure to 10^{-10} torr, rather than up to atmospheric pressure? What effect results when you continue to go up in pressure once you are above that required for monolayer coverage?

P. G. Shewmon (Carnegie Institute of Technology): Let's start on the low-pressure end first; then we'll talk about the high-pressure end. I think that the problem with going to low pressures is that you have more evaporation. If you have a gas at 1 atm over a metal, although the metal has a finite vapor pressure, the gas tends to reflect the metal atoms back, and you do not have as much loss by vaporization. If you go to a really good vacuum, you could have some vaporization problems. However, I know of some people who have worked with field emission tips of copper and gotten away with it. The problem of working in high vacuum is not insurmountable; I think it's more a matter of nobody getting around to it yet.

On the high-pressure end, we actually did some work at 0.1, 1, and 10 atm, theorizing that maybe this long mean free path is really some kind of partial evaporation and bouncing back. However, we could find no effect of pressure at all in the range of the order of about 1 atm. Thus, I don't think going to higher pressures would have any effect at all until you got to extremely high pressures and got into activation volume problems or something like that.

G. W. Rengsdorff (Battelle Memorial Institute): This is a comment rather than a question. Both you and Prof. Birchenall reported evidence that atmosphere does not much affect the surface diffusion of copper. I would like to present evidence that this is not true for iron. It is based on observations of grain boundary growth during floating-zone purification of bars of ½- and 1-in. diameter. Active-gas atmospheres are used to boost zone purification by the use of chemical reaction. On the first and third zone passes where hydrogen is used, the grain structure (at least the high-temperature grain boundaries) is grooved enough to be visible. On the second pass where argon is the atmosphere, the grains appear sharp and clear. On the fourth pass where diffusion-purified hydrogen is used, so little grooving has occurred that the grain structure cannot be seen at all. Do you have any comment?

P. G. Shewmon: I don't think that either of us could deny that your active gases have a very profound effect on the surface diffusion coefficient, but this is my only comment.

L. J. Bonis: I would like Prof. Shewmon to clear up some of the differences of opinion he has with Prof. Birchenall.

213

P. G. Shewmon: I don't really know that we have any differences of opinion. I think I differ in what you might call degree of optimism or degree of naïvete. I would say I'm more optimistic, but Prof. Birchenall might say I'm more naïve. One area of what you might call reasonably strong disagreement concerns the fact that I don't like to use activation energies that are obtained by tracer techniques or compare them with those obtained by mass transfer for the following reason. If volume diffusion is completely ignored in the tracer work, then we're simply talking about completely different things. If you insist on talking about the older tracer work, then I think you must compensate for the volume diffusion somehow. Until something like that is done, I think it is extremely misleading to even talk about these things together.

J. M. Williams (University of Delaware): My research involves a surface-diffusion study on iron by a tracer technique. Thus, we need to account for the losses into the volume, if indeed in a certain temperature range this is significant. I believe we have a suitable solution to the mathematics for such a situation, and the published results will show the effect of this parameter δ. Somehow the value obtained seems rather large, for reasons unknown at present. The Russian literature wherein the mathematics derived by Geguzin[*] are used (which mathematics, in my opinion, may be incorrect, but nevertheless may be a significant approximation even if it fails to give the precisely correct value) reports for a variety of systems values of δ, the thickness of this highly mobile surface layer (or at least an effective thickness of it), of $\sim 10^2$–10^4 atomic diameters. Intuitively, this seems unreasonable for any sort of usual crystal structure at the surface. Indeed MacRae and Germer[†] have shown in experiments with iow-energy electron diffraction on nickel that there is no such tremendous reorientation of atoms in the lattice near a surface. Nevertheless, perhaps there is a finite thickness—more than 1 or 2 atomic diameters—to this region. The work of Drew and Pye[‡] on silver is the only work which shows a decrease in the observed surface diffusivity at higher temperatures. I have treated their data with a mathematical model in which this thickness δ was a disposable parameter and have determined what values would account for the deviations which they report. (They report deviations from an Arrhenius-type relationship.) The values obtained for this particular parameter are indeed rather large (10^3 or 10^4). At the moment, I don't know what it really means, but I think the answers to these questions must await further and better experimental results.

P. G. Shewmon: In connection with this δ, what we do is arbitrarily assume that we should take the atomic diameter to be δ. We have reported all our diffusion data in this way. In fact, what you can measure out of the analysis that Mullins[§] gave or the tracer analysis that I've given is the product of the surface diffusivity and this thickness δ. We have no more conditions with which we can evaluate D_s and δ separately. If the Russians do get separate values for these two quantities,

* Ya. E. Geguzin, G. N. Kovalev, and A. M. Ratner, *Phys. Metals Metallog.* *(USSR) (English Transl.)* **10**: 45 (1960).
† A. U. MacRae and L. H. Germer, *Ann. N.Y. Acad. Sci.* **101**: 627 (1963).
‡ J. B. Drew and J. J. Pye, *Trans. Met. Soc. AIME* **227**: 99 (1963).
§ W. W. Mullins, *J. Appl. Phys.* **28**: 335 (1957).

they must have used additional physical assumptions, thus enabling them to make separate evaluations. Can you tell me what additional physical information they have used that we haven't?

J. M. Williams: This comes about by their manipulation of the experimental concentration distribution, which is appropriately plotted *versus* distance to the $\frac{4}{3}$ power, yielding a curve of a certain shape. Actually, I guess I could best explain this by writing down their mathematics. What they have done is to obtain an expression for what they call $\bar{c}$, which is the integrated concentration. Then, they have used the sectioning technique to measure $\bar{c}$. They have taken slices normal to the surface and counted the radioactivity in each slice; $\bar{c}$ is the activity in the slice. They have obtained an approximate expression of the form

$$\bar{c} = A \exp(-BX^{4/3} - CX^2 - DX^{8/3} \cdots)$$

where A, B, C, and D are various combinations of the parameters of the problem and X is the distance from the source. B is an involved conglomeration of the various parameters of the problem, but C is a much simpler expression. Now, if one plots $\ln \bar{c}$ *versus* distance to the $\frac{4}{3}$ power, then the slope of the resulting line gives B at $X = 0$. The product of the thickness and the surface diffusivity is contained in B. C contains only the surface diffusivity. A plot of the sum $(\ln \bar{c} + BX^{4/3})$ *versus* X^2 yields C. Since C is a function of the surface diffusivity only, they can determine both the surface diffusivity and the product of D_s and δ, which comes from B.

To comment on this, their experimental data for iron do this remarkably well, and they have other systems that also exhibit such behavior, and, consequently, they obtain good values. However, as we mentioned, Pavlov and Panteleev* have pointed out that the experimental conditions which they used in the determination for iron did not satisfy one of the assumptions that was necessary to obtain the approximate form to start the whole procedure. Therefore, this does indeed cast doubt on their results. What I don't understand is how such a complicated analysis of something as hard to obtain as these concentrations fits so well. There's a good deal of error involved in getting these $\bar{c}$'s. I don't want to accuse the Russians of doing anything unorthodox, but I'm dubious.

L. J. Bonis: Prof. Blakely, do you have any comments on the papers?

J. M. Blakely (Cornell University): Prof. Birchenall mentioned that what you actually measure in the mass-transport experiment is the product of the diffusion coefficient and the surface free energy and that there is, in fact, some doubt about what to use for the surface free energy. I should point out that, if you work in a range of dimensions such that volume diffusion is the process that's causing the transfer, then the corresponding product is the volume diffusion coefficient and the surface energy. By measuring the rate of mass transport by volume diffusion, you get a value for the surface free energy which is the appropriate value to use in interpreting the transport by surface diffusion. The other point that I'd like

---

* P. V. Pavlov and V. A. Panteleev, *Soviet Phys. Solid State (English Transl.)* **6**: 955 (1964).

to mention is that the surface diffusion coefficients we've been talking about will not, in general, be those that apply in crystal growth from the vapor. In that case, the atoms are deposited on surfaces that generally are very close-packed planes, and so we certainly don't have the requirement of creating adsorbed atoms. The surface diffusion activation energy that would be measured there is simply the activation energy for motion of the adsorbed atoms and doesn't involve the creation energy.

L. J. Bonis: Dr. Li, do you have any comments?

C. Y. Li (Cornell University): I would like to say something about surface diffusion. I feel that, at the present, we have limited knowledge of surface structures, especially the defect structure on crystal surfaces. For simple systems, such as the ionic crystals which we have discussed in our paper, we really do not know the exact vacancy concentrations at the surface. In our calculation, we have neglected the vertical lattice distortions and the induced electronic dipoles at the surface, and we have assumed that near the crystal surface the dielectric properties are the same as those in the bulk crystal. An understanding of the arrangements of adsorbed ions on ionic crystal surfaces is also lacking. The situation on metallic systems is far from desirable. We can find out long-range atomic arrangement on crystal surfaces by doing low-energy electron diffraction experiments, but we cannot say anything about the defect structure. More experimental and theoretical work is required in order to understand the defect structure on crystal surfaces before we can propose any reasonable surface diffusion mechanism.

L. J. Bonis: The following is a question I'd like to ask all the authors. What do you think, in your field, is the most pressing problem that needs to be worked out in the future? What do you feel, in your field, is the most important answer we need to obtain?

J. Turkevich (Princeton University): I'd like to comment on Dr. Li's remarks. I think that as we review chemical kinetics and reactivity, we're coming to the realization that an extrapolation from the structure of molecules such as it exists in the equilibrium state to the state of the structure at the point of reaction is one big, big extrapolation. The problem in kinetics is to derive methods to determine what a molecule looks like when it reacts (when the structure has a 10^{-8} or 10^{-12}-sec lifetime) from its known structure when it has an infinite lifetime. For catalysis and other systems as well, we need more dynamic measurements and studies of transients. We must nudge the systems a little to see how they respond. I envision this type of investigation for physical chemistry during the next decade.

L. J. Bonis: Dr. Li, what do you think is the basic cause of our inability to determine the structure of the surface? Is this the result of the limitations of our tools, or a lack of understanding of basic phenomena?

C. Y. Li: It's both. Experimentally, as I mentioned before, low-energy electron diffraction allows us to look at the long-range atomic arrangement on crystal surfaces. Some observations have been made using the electron microscope to observe surface steps by decorating the surface. We have to develop techniques so that the defect structure on crystal surfaces can be revealed. Theoretically, at the present, we do not know how to characterize the potential distribution at metallic crystal surfaces. Some work has been done by assuming certain arbitrary potential distributions. As a consequence, we cannot satisfactorily calculate the formation energies of vacancies on metal surfaces or the saddle-point energy for atomic jumps.

P. G. Shewmon: Actually, the field-ion microscope allows us to get down and look at atoms, doesn't it?

C. Y. Li: The metal surface is not at thermal equilibrium due to the existence of a "field evaporation" process.

J. Turkevich: The situation is somewhat like trying to analyze water while standing on Niagara Falls.

C. E. Birchenall (University of Delaware): I think that there isn't much reason to despair. I can remember twenty years ago the problems related to volume self-diffusion in metals were very similar to those for surface self-diffusion now, and I believe that, by studying many different approaches to the problem, we finally settled on a few that worked very well. The results from different laboratories now agree, especially on silver (everybody does silver). We ought to declare a moratorium on publishing volume self-diffusion measurements on silver. I think that the surface self-diffusion problems are going to work out in exactly the same way; we'll add a little bit of information at a time, and in ten or fifteen years we'll agree pretty well on how to go about it properly.

Now this, to me, is the foundation for asking about the more complex problem of how foreign atoms move on metallic surfaces and on other surfaces of simple crystalline substances; this is going to be the eventual basis for explaining many of the phenomena that are observed in the very complex systems that are used in catalysis. This doesn't mean that anybody whose livelihood depends on the use of catalysts is going to wait until we work out all these things. They will continue to develop working catalysts and processes and set up some of the problems that may prove to be keys to that kind of investigation where the fundamentals exist on which to build a theory. So I think that we're in a reasonable state and that things aren't desperate by any means, and if we proceed in the way we're going, we'll get the right answers.

L. J. Bonis: It strikes me that there are many types of surfaces. There are many questions and models I have seen just in the past five or six years concerning close observations of surfaces and defects, and yet we're actually complaining that we don't have enough methods and close observation. Do we really not know what a surface is and do we have the ultimate in these models?

C. E. Birchenall: I'd like to point out the methods of observation and the resolution that they present, as Dr. Turkevich mentioned, have improved greatly in recent years. We're getting closer and closer all the time, so this is not a desperate situation. It's a very optimistic one. If all these problems evaporate, we'll have to go out and comb beaches.

J. Turkevich: Of course, on the surfaces, one of the trends has been to go to materials such as molecular sieves, coordination complexes, and the like, where many of the problems of atom diffusion are not removed, but minimized, and I think that these systems have surfaces that are more sophisticated than simple metal surfaces or metal oxide surfaces.

L. J. Bonis: For the last decade or so, what was the most helpful tool for studying surfaces? Was it the ultrahigh vacuum or the microscope? What actually gave a boost to the studies? Obviously, it has been largely in the last ten years that most of the advances were made.

C. E. Birchenall: The thing that gave the biggest boost was the need to know.

J. Turkevich: Low-energy electron diffraction is certainly giving valuable information about surfaces. It's affecting the surface less than any other tool. On the other hand, it may be that in preparing such a surface for examination, we are making an artificial surface, a surface that a chemist normally doesn't deal with. We have either a physicist's surface that exists only 10^{-3} sec in an ultrahigh vacuum and is contaminated very rapidly, but is well-defined, or a chemist's surface that we cannot define, but which is stable and very useful. Molecular sieves, I think, come closer to realization of a synthetic type of a surface that is well-defined, stable, and useful. Another well-defined, useful surface may be produced by adsorbed complexes of known structure on surfaces.

V. De Palma (Cornell Aeronautical): We've been talking about surfaces which we're accustomed to seeing. What about a surface that results from a vacancy cluster on the inside of a material, sort of a piece of void on the inside? That's a true surface, isn't it? And, if so, will the property ascribed to the outer surface to which we're accustomed be attributed to the properties of the internal surface?

J. M. Blakely: The use of macroscopic quantities, such as surface free energy, in the description of a vacancy in a metal was first suggested by Brooks.* With use of this concept, a rough estimate of the vacancy formation energy can be made. However, the correspondence should not be taken too literally, since the "surface" of a vacancy or vacancy cluster is one of extremely high effective curvature and involves only a small number of atoms. Thus, in a metal, for example, the electron distribution into the vacancy does not necessarily correspond to that at a free surface of the crystal.

E. Mendel (Geoscience Instruments Corp.): I hope I'm not being too naïve, but is there anybody on the panel who would care to define the type of surface we're dealing with?

P. G. Shewmon: We all agree that it involves a close-packed plane at absolute zero. If you deviate slightly from this and talk about a surface that cuts through the crystal at an angle to this close-packed plane, you have closed-packed planes ending on the surface, and then these are shown as steps. However, these steps can also disorder, and kinks or jogs probably will form in these. This sort of a thing is probably valid, or undoubtedly valid on a clean surface at very low temperatures, and most of us hope that it is valid at quite high temperatures where there is a good deal more thermal disorder and where we still hope we don't have our surface completely cruded up by adsorbed impurities residing at these

---

* H. Brooks, "Impurities and Imperfections," ASM Seminar, 1955.

quite high temperatures. I know of relatively little information on actual structures at high temperatures.

E. Mendel: Then you conceive of the surface as being a two-dimensional entity, rather than a three-dimensional one. For instance, some people, when they speak of surface preparation, speak of the damaged layer. Now at this point, they're no longer speaking of a two-dimensional entity, but rather of a three-dimensional entity. I'm sure that two people getting together in a meeting such as this might be speaking two different languages—each having his own definition of a surface. I pose this question in order to bring out this fine point.

P. G. Shewmon: Unquestionably, there is damage resulting from the method of producing this crystal or this surface. At least in high-temperature work, you work near the melting point, and a great deal of this material should anneal out. Also, if you transport appreciable quantities of matter from one part to another, you're taking what damage was there and either removing all the atoms by eating out that part of the crystal or piling perfect crystal (or at least other crystal) on top of it. If you're talking about surfaces that are used without this high-temperature anneal, then certainly garbage of one sort or another or damage would introduce a third dimension, or, as I prefer to say, this will influence the properties of the surface of which I spoke.

S. Bowers (Fairchild Publications): What are some conclusions regarding changes in aerospace materials as they enter outer space? Which aerospace materials are most subject to change? What is the nature and suggested remedies?

I. R. Kramer (Martin Company): The change in the mechanical behavior depends upon the application as well as the material. For structures where creep is involved, we expect that all metals will have an increased creep rate in the low-pressure environment. The low pressures will, as shown, increase the fatigue life, and, therefore, design parameters based on atmospheric conditions will be more than sufficient for space application. Eventually, it may be possible to take advantage of the improved fatigue life to design a lighter-weight structure. With regard to the suggested remedies, I think that all of the problems may be taken care of by appropriate design.

V. De Palma: In the ion-bombardment cleaning of the aluminum wire, what was the energy of the bombarding ion? Could the resulting decrease in plastic flow be due to the presence of argon ions in the aluminum lattice? What were the dislocation densities for the aluminum crystal in the experiment?

I. R. Kramer: The potential between the cathodic specimen and the anode was 400–600 V. An ion current of 0.3 mA/cm^2 was used. No doubt the impingement of the argon ions on the high-purity aluminum caused damage—probably by embedding argon in the lattice. The dislocation density was not measured.

L. J. Bonis: I notice that your data indicate very little difference between testing at 10^{-7} torr and 760 mm Hg. Isn't that so because the monolayer formation for 10^{-7} torr is still inside of a few seconds?

I. R. Kramer: No, the effect was rather large.

L. J. Bonis: Yes, but you had a value of 1.2×10^{-7}.

I. R. Kramer: Oh, you mean between 1.2×10^{-7} and 8×10^{-7}. That's within the scatter of the stress–strain curve.

J. J. Duga (Battelle Memorial Institute): I have one comment that I want to address to Dr. Kramer. In regard to the observed surface dependence on creep, for example, it seems that several important points have been glossed over and dismissed in a rather cursory manner. If one assumes that plastic flow or the lack thereof is even vaguely related to dislocation motions and interactions, then it seems most reasonable to assume that the diffusion of foreign adatoms into a crystal will have a direct influence on such dislocation motion. Furthermore, the presence of such diffused impurities can influence the local stoichiometry, so that one can get an indirect influence on dislocation motion. Conversely, the physical and chemical nature of the material under investigation can have a marked influence on the backward diffusion of extreme impurities, which should also affect this dislocation motion. How can you reconcile these observations which have been fairly well-documented, particularly in the work of Kröger and Vink* on stoichiometry in binary compounds, with your own views on the all-importance of surface conditions with their contributions toward creep and other mechanical properties?

I. R. Kramer: In these experiments, dislocation climb by vacancies is not important, and, therefore, impurity diffusion is not important. In any case, even when impurity atoms affect the creep behavior, surface effects are still important. The change in the creep rate due to the accumulation of dislocations in the surface layer would be superimposed on the changes due to impurities.

J. Turkevich: Dr. Kramer, is there such a thing as the Rehbinder effect?

I. R. Kramer: Yes. We have investigated the Rehbinder effect.

J. Turkevich: But it's not what he claims.

I. R. Kramer: The explanation Rehbinder put forth is based on the presence of a monolayer, and this theory would not give a bell-shaped curve. We have found that the surface-active agent behaves in a manner similar to that in the electropolishing experiments. In other words, it's the rate of removal of the metal that is important. The reaction between the surface-active agent and the metal is such that a metal soap is formed. The rate of solution of the metal soap will control the change in the mechanical behavior.

J. Turkevich: I'm interested in it not from the metallurgical point of view, but from the viewpoint of a student of Russian chemistry. There have been all sorts of claims made by Rehbinder, and the Soviets think that is going to solve all of the problems of Soviet metallurgy.

I. R. Kramer: I was going to say the effect is not a very strong one. For example, we could not find any effect on polycrystalline material. The Russians do report an effect on fatigue; however, we have not investigated this effect.

J. Turkevich: Well, as I recall, the British hit Rehbinder pretty heavily in an article in *Nature*.

---

* F. A. Kröger and H. J. Vink, in: *Solid State Physics*, Volume 3, F. Seitz and D. Turnbull (eds.), Academic Press (New York), 1956, p. 307.

I. R. Kramer: Yes, this refers to Cottrell's* work, which shows you must have the oxide present. The basis for this is that a metal soap is not formed unless oxygen or water vapor is present. However, in some fairly recent work, Hilton Smith claims that during deformation exo-electrons will supply the necessary energy to cause the formation of the soap. However, we could not find a Rehbinder effect on gold.

F. Satkiewicz: Can anyone comment on the practical problem of dust in a vacuum system? For example, there must be a finite probability that a piece of organic dust will fall on your surface, and momentarily you will have a situation where you've changed the state of the system. Do experiments take this kind of a thing into account? Is this effect significant?

J. Turkevich: Not particularly, since our experiments can be correlated with the physical properties of the ultrafine particles. We certainly try to avoid this, but there are other factors involved. Bursts of cosmic rays may enter, which produces ionization and nucleation.

F. Satkiewicz: Well, in this case, we're talking about surface diffusion and perhaps the introduction of a carbonaceous species on the surface.

J. Turkevich: I find that dust usually comes in big particles. We do not find dust in the 100- or 10-Å ranges.

F. Satkiewicz: That makes it all the worse.

J. Turkevich: The large particle just sits there in one big piece, and that's all it does.

F. Satkiewicz: At 800°C?

J. Turkevich: I imagine so, if it is inorganic. If it is organic and in a vacuum, it will volatilize from some hot material to some cold spot on the bell jar.

F. Satkiewicz: But in the interim, what effect does this have on your measurements?

J. Turkevich: We can't tell. If you worry about these things, you just don't start the experiment.

F. Satkiewicz: But you worry about the small concentrations of vapor.

J. Turkevich: Very true.

L. J. Bonis: You have to worry about some things.

J. Turkevich: These are controllable things; these are theoretically amenable.

L. van Someren (Thermo-Electron Engineering Corp.): Going back to this business of surfaces at very high temperatures, what do they actually look like? Does anybody see any chance of actually finding out or answering the question at all? In other words, electron diffraction enables us to look at the surface over an area which is large compared to the size of an atom and gives us results which reflect averaging over this area. The atoms at elevated temperatures are going to be moving around a bit; their positions are not going to be well-defined. This seems to me to prevent the use of low-energy electron diffraction on surfaces near the melting point. The only thing that I can think of that might possibly be

---

* A. H. Cottrell and D. F. Gibbons, *Nature* **162** : 488 (1948).

of any use under these conditions is grazing incidence electron microscopy on the actual specimen surface, but then you have problems with getting the specimen near the melting point. Does anyone have any comment on how to determine the structure at high temperatures?

J. Turkevich: I think the approach of electron microscopy is certainly a very good one in that the signal-to-noise ratio is very great. You can see the beginnings of things. If you have a surface that's just beginning to get disrupted, you can, for example, take a material and gradually bring it up to the melting point or as close to it as possible. You can take replicas or do any one of various stunts and see what happens and gradually approach the situation where the whole surface collapses. Another approach to surface mobility is the magnetic resonance–electron-spin resonance approach. With X-ray and diffraction phenomena, you have to have an *ensemble* of maybe 100 or 200 atoms that are forming a pattern, and you can get any kind of information about it. Magnetic resonance deals with *individual* nuclei. In studying manganese–manganese interaction in electron-spin resonance, one can see whether the manganese atoms are moving away from each other or clustering; in studying solid-state and surface transformations, the various magnetic resonances offer a very strong tool. One can detect 10^9 species with spin resonance. The double-resonance technique can tell us whether one atom is close to another atom, although they may not be ordered with respect to one another. I agree with Prof. Birchenall; just give us a little time and we and our successors will solve these problems.

Index